Copyright Anne Hamilton
June 2014
Seventeen Mile Rocks
QLD 4075 Australia

National Library of Australia Cataloguing-in-Publication entry
Author: Hamilton, Anne, 1954- author.
Title: Gawain and the four daughters of God : the testimony of
          mathematics in Cotton Nero A.x / Anne Hamilton.
ISBN: 9780980362060 (hardback)
Subjects: English poetry--Middle English, 1100-1500.
          Mathematics and literature.
Also Titled: Four daughters of God
          Gawain and the Grene Knight.
          Patience (Middle English poem)
          Pearl (Middle English Poem)
          Purity (Middle English poem)
Dewey Number: 821.1

Gawain and the Four Daughters of God:

the testimony of mathematics in Cotton Nero A.x

Anne Hamilton

Gawain and the Four Daughters of God:

the testimony of mathematics in Cotton Nero A.x

Anne Hamilton

Contents

introduction

To venture into medieval poetry using a mathematical spotlight is not a unique approach. Still it is sufficiently uncommon to require some explanation and—perhaps—defence. Whether it also requires an apology I'm not entirely certain.

Paul Davies in *The Goldilocks Enigma* mentions that, long before Einstein, Augustine of Hippo referred to the concept of relativity.[1] As far as Davies is concerned, nothing of significance can be attributed to this insight because Augustine failed to accompany it with any sort of equation. Davies considers that, given the human imagination and its ability to conceive outlandish ideas, it would in many ways be astonishing if someone in the past *didn't* come up the notion. So, without a definitive equation, he rejects the thought Augustine was a genuine forerunner of Einstein. I'd like to think that Davies, a theoretical physicist, was not overreaching himself in so casually dismissing the learning of the past. But I'm not entirely sure. Because the moment I saw his comment, I asked myself: *what would an ancient equation look like?*

This question doesn't seem to occur to Davies who appears completely unaware words were once alphanumeric in nature. And that, ironically, something of this ancient fusion of letters and numbers is preserved in modern algebra which uses letters as generic substitutions for numbers.

The possibility an undiscovered equation is embedded in Augustine's writings might be improbable but it is certainly not impossible.

However, while such a thought as this—that there is mathematical encoding in the text—is far less mind-warping than relativity itself, it is in fact generally considered to be beyond the fantastic. To combine words and numbers, so that sentences become equations in their own right, makes mathematics suddenly more like fantasy than the 'handmaid of science'.

To take such a possibility into the world of ancient literature risks entering a world as strange and faery-like as Tolkien's Mirkwood and seeing neither the wood nor the trees. Just the unfamiliar.

Yet, as in that fabled dark forest, there are occasional well-marked paths through its reputedly trackless wilderness of thorn and mist and briar.

Literary mathematics may hold its secrets close to its heart. However by venturing close enough, it is possible to interpret the strange gleaming sigils which can be glimpsed from time to time, making and unmaking themselves.

It's not just imagination: there really *is* something going on in the deeps of the wood. We can be certain of that because there's one sure thing which can be said regarding mathematics. Its results are *testable*.

'Testable' means that, if a particular pattern suggests a certain equation, it's possible to check for it by predicting additional features of the text. This has been my approach. I've used the basic scientific method—develop a hypothesis, then test it repeatedly.

When it comes to the attitude of critics toward arithmetic and geometric features in medieval poetry, there is a tendency towards extremes. In the middle is the vast majority: the reviewers who ignore any mathematics uniformly and entirely.

Of those commentators who do notice it, most simply write off the phenomenon as numerological in nature. The unfortunate consequence of such an automatic labelling is that mathematicians are dissuaded from having a closer look. A handful of scholars steadfastly make the point numerical literary design is radically different to numerology. However, their findings are apt to be quoted as examples of numerology anyway, regardless of their protests to the contrary.

To suggest numerical literary design automatically equates to numerology is as logical a proposition as insisting all verse is magical incantation. Just as a poem is not necessarily a ritual spell, neither is arithmetic metaphor necessarily numerology.

The marriage of mathematics and literature lasted at least two and a half thousand years, quite possibly much longer. Buddhist scriptures are said to be as much temples of mathematics as of words. The New Testament is a wordscape with mathematical blazes indicating trails through it back into the Hebrew scriptures.

Greek literature from at least the time of Plato demonstrated an adherence to classical ideals of elegance in form and structure. 'In the *Timaeus*, Plato (427–347 BC) used Pythagoras' so-called perfect numbers[2] as a means of describing the perfection of the natural harmony that existed in the world and universe. He combined language and number to define what he called the measure of 'the world soul'...'[3]

Even before Plato, there is evidence of the use of perfect numbers in the prophecy of Isaiah.[4] The Christian authors of the gospels and epistles retained this widespread idea of beauty and perfection in composition, developing their own distinctive 'numerical literary style'[5] whose last faint echoes dissipate at some time during the Renaissance.[6]

'We know relatively little about the medieval view of the poem as a "well-wrought urn" in which artificial shaping is as important as the clay'[7] is a common sentiment today, reflecting the tendency to believe that the comparatively recent separation of mathematics and literature is how it always was.

Yet CS Lewis noted the comment of Albertus Magnus who linked *boni imaginativi* with a talent for mathematics[8] while Richard Schoek flung down a gauntlet with his reminder of the words of Hugh of St. Victor's twelfth–century *Didascalicon*: 'mathematics never operates without the imagination.' Schoek recalled, amongst many others, Dante in *De Vulgari Eloquentia* describing the structuring of a stanza and *canzone*, Gerard Manley Hopkins expounding the equation of a sonnet and Ezra Pound stating that poetry is inspired mathematics.[9]

It might also be noted Omar Khayyam was such an outstanding mathematician he completely solved the general equation of the third degree. Robert Edwards recalls that, in *The Squire's Tale*, Chaucer introduces a marvellous mirror that 'incorporates mathematical formulas in order to reveal the subjective truth of language.'[10]

As far as the Middle Ages goes, if anyone is recognised as using serious mathematical design, it's usually Chaucer. However he said to be an 'anomalous case', familiar with arithmetic from his use of the counting board in the Customs House.[11] But is he?

Even if there were no testimony from Chaucer or the other poets, the mathematics itself speaks. It shouts. It draws attention to itself with shameless brazen posturing. I cannot subscribe to Peterson's restrictive views that an unambiguous statement from the poet is necessary before it is possible to suggest a text has deliberate mathematical design. Consider for example, the stanza lengths of *Summer Sunday*—13, 13, 13, 13, 13, 13, **8**, 13, **8**, 13 and 13. Those sudden changes to eight lines are simply flagrant self–advertisement for an underlying mathematical design based on the golden section.

Unlike numerology which veils its meaning and seeks obscurity so that only the initiate will comprehend the hidden form, mathematical metaphor tries its hardest to be noticed. This flaunting often takes the form of a sudden irregular stanza which, in many cases, is attributed to scribal error. Sometimes the metaphor sits against an illuminated letter, its position noted by scholars time and again but not its significance.

Unfortunately, the widespread assumption there is no numerical

design underpinning the majority of texts has sometimes led editors to purge their editions of 'scribal errors', to normalise stanza lengths and to make irregularities invisible. This obscures the design even more greatly. On several occasions, I was able to retrieve a long-discarded hypothesis on discovering a line I'd predicted would be marked in significant way had been rectified by an entire series of over–zealous editors.[12]

It's one thing to be baffled by mathematical metaphor. It's another to decide it doesn't exist. Jeffrey Burton Russell tells the story of a scholar, well–versed in medieval literature, who was completely perplexed by a sudden brief mention of a 'burning bush' which seemed completely out of context.[13] This cultural chasm widens almost immeasurably when it comes to literature and mathematics.

A rose is a rose but also not a rose: it is one of the most obvious symbols in poetry. We don't mistake a symbol for a spell in medieval love tokens. However suggesting 1111 is a mathematical metaphor or 490 is a numerical motif but that neither is numerological in nature far too often seems to ask for too subtle a leap of understanding. Numerical tokens within medieval poetry do not have to have occult magical overtones. It is my belief that part of the failure to recognise the difference between numeracy and numerology in poetry is the result of an inability to recognise the long–standing animosity between neo–Platonism and Christianity.

The early Church Fathers, in particular Augustine of Hippo and Gregory of Nyssa, may have been sympathetic to neo–Platonism but that was the fourth century. Too often it has been assumed their influence extended all the way through the Middle Ages—that Christianity was virtually monolithic in its views across a bridge of a thousand years. This ignores the natural ebb and flow of culture as it mixes with spirituality. It takes no account of frequent influx of heresy or of the influence of Celtic monasticism on the continent. It discounts the political manoeuvrings resulting in such diverse events as the proclamation of Charlemagne as Holy Roman Emperor and the election of an anti–pope in the late fourteen century as having any impact on clerical thinking.

Most importantly for this analysis it ignores the profound challenges posed by twelfth century neo–Platonism which fused courtly love and occult mathematics with the poetic impulse— resulting in a 'weaving of words' only marginally removed from incantation.

Not every use of the golden ratio was neo–Platonic in inspiration. Sometimes exactly the opposite was the case.

Until the late nineteenth century, mathematics was akin to experimental science: it had not elevated the ancient Greek concept of 'proof' to Olympus and deified it. Paul Davies' call for a relativistic equation from Augustine belongs to this mental shift.

Certainly there were individuals who were rolling the idea of 'theoretical proof' around the lower slopes where a seachange was already in the air. Paul Gordan, confronted with the revolutionary work of David Hilbert, declared, 'This is not mathematics. This is theology.' Hilbert's vision of the beauty and harmony of numbers may have seemed ethereal at the time but is now readily accepted by mathematicians who now recognise an enigmatic connection between symphonic music and prime numbers.[14] Ironically, despite Gordan's comment, the separation of mathematics and theology has been formalised for centuries. It seems to have coincided with the divorce of mathematics and literature, in fact.

This has produced a curious state of affairs: there is possibly no one around today who, despite our enormously greater knowledge, is by medieval standards more than *semi*–literate. Our knowledge of the trivium (grammar, logic and rhetoric) is usually in inverse proportion to our knowledge of the quadrivium (arithmetic, geometry, astronomy and music); in fact, generally speaking, the deeper our knowledge of any one particular discipline, the more superficial it is of any other.

In saying this, I introduce my qualifications for writing this book which is, admittedly, a curious hybrid, as much a quest story as an academic analysis. To follow the mathematics it is necessary to forgo the rigid strictures of a closed discipline and to think like medieval man. To be frank, restraining my exuberant joy at the discovery of some truly nifty mathematics in favour of a detached scholarly tone is not what I feel is appropriate to a celebration of artistic genius.

Despite my degree in mathematics and my decades of teaching it, I am only semi–literate. I love pure mathematics in its airy loftiness, but I also love the way medieval poets wove mathematics through all they did. It is worth noting that, although I have explored using a mathematician's eye, I have not taken the view that any medieval poet used mathematics for its own sake.[15]

That idea, which seems prevalent in most current analyses, is a post–modern variation on 'art for art's sake'. As I sit writing and listening to an amazing composition based on the DNA sequence of starfish, I can't help but think how remote I am from those times: what is a stunning musical concept to me would have been so obvious to medieval men and women. I don't have an adequate knowledge of the quadrivium, let alone one that encompasses the trivium. Despite the fact I've taught geometry from time to time, my knowledge of it would have appalled a medieval schoolboy.

Yet as I have trespassed into medieval studies and blundered my way through unfamiliar terrain, I have been astonished by the gracious hospitality I have encountered and the willingness of various scholars to answer my awkward, half–formed questions. I must particularly single out Ed Condren, Robert Stevick and Keith Atkinson, but I am also greatly indebted to Joan Helm and David Howlett.

There are several reasons for choosing the poems of the *Pearl* manuscript as my formal garden of exploration. The first is the modern concept of 'proof'. Here is a sufficient body of work to create a hypothesis *and test it empirically*.

I will not be examining numbers like 5 or 3, 7 or 1 which form the backbone of many numerological readings. Of course they will form part of the analysis, however my emphasis will be on numbers like 1743, 81, 1056, 1111, 490, 153 and 231. These are numbers which are not amenable to just any interpretation, but only to very specific ones.

Secondly, the *Pearl* poet is a mathematician in the truest sense— his work is intricate, but it exudes an air of consummate elegance and delicate grace. There is a simplicity to his work at times as well as unexpected playful depths. He's a generous soul, too: not only has he marked his trail well but he had also left constant confirmation for anyone following the path.

Perhaps most surprising of all, however, I find a transcendent beauty in his mathematics. His arithmetic metaphors are a source of unutterable wonder. His clues to the meaning of his work are obvious, are repeatedly corroborated and are methodically laid-out. He isn't secretive: his mathematics is very open. However it is also very important to him; so important that it is debatable whether it is possible to understand the poems without examining the mathematics.

At least today.

In the fourteenth century, a general reader would probably not have needed any mathematical analysis to have recognised the overall theme of the *Pearl* manuscript. These days, however, we need every hint we can get.

My third reason for choosing the poems of the *Pearl* manuscript is because of their influence. It seems very likely there was never more than one copy of the manuscript. If such was the case, then it almost certainly made no impact whatsoever at the time it was written. Yet, the writings of two twentieth century authors—both of whom had books in the 'top ten' most popular works of the century[16]—show how deeply it influenced their own thinking. It is a work that has come into its own centuries after it was first composed.

The fourth reason for choosing the *Pearl* manuscript is that the poet has such wide-ranging interests. Sometimes in his poetic cathedral, I just want to point and say like a gawking tourist, 'Wow! Is that a labyrinth? Hey, isn't that "rose window" fantastic? See this knotwork—is it the same as that over there?' At other times I want to stop entirely and delve into a detailed analysis as well as a comparison with other authors who share his same enthusiasm for this fusion of numbers and words. This will mean at times casting a glance back at the Anglo-Saxon poetry of the dark ages and beyond it to the New Testament, while at other times noticing the remarkable resemblance between the mathematics of the *Pearl* poet and certain iconic works of Leonardo da Vinci.

There is a widespread assumption by historians the medieval era was one of the great 'Ages of Faith' in which 'Christianity was reckoned to be a public truth'[17] and when it waxed strong as the church triumphant throughout most of Western Europe. However, as we shall see throughout this investigation, this is at best an extremely simplistic view. Indeed if it were true, then we would not have the *Pearl* manuscript in the form it exists. It is, at least in part, a call to battle.

It will be difficult to do justice to the poet's work without at times making it seem more and more complicated. On occasions, the technical aspects of 'folding', 'infolding' and 'unfolding'—terms used by Renaissance critics and rhetoricians to describe the 'convoluted, enigma–making processes of allegorical poetry and exegesis'[18]—may seem truly overwhelming. The soaring architecture is dominated, however, by a handful of numbers chosen with obvious care for their theological meaning. The mathematics is basic.

In this cathedral of mathematics, there are side–chapels and galleries resplendent with numerical and geometric tapestries but nothing is there simply because it looks good. It's there for a purpose. I will explore some of these places while others, still in deep shadow as to the intention behind the mathematics, will have to be left for the moment.

There is much about the poems of the *Pearl* manuscript that is typically medieval in expression and conception, but there is much too, particularly in *Sir Gawain and the Green Knight*, that seems original. If it is not, at least it is covered with a highly artistic render. In the same way, both the commonplace and the original appear in the mathematics.

A single brilliant mathematical idea dominates the overall structure and appears to be so stunningly creative that, if it's not the poet's own conception, then really serious issues arise about our understanding of number theory in the fourteenth century. Fairly serious issues are going to arise anyway, but ultimately we will be faced by the choice of a single towering genius whose poems are a unity of groundbreaking mathematical elegance or else a master of arithmetic, geometry and music who drew on a body of prior learning which has either been lost or not uncovered yet.

Whether or not the *Pearl* poet was a literary genius, I will let others judge who are far better qualified than I am. Take it from me, however, he was an extraordinarily gifted mathematician. His number work is visionary, even if parts of it are derivative; his is a finely-crafted artwork of astonishing scope and creativity. The rare genius of it comes down to its simplicity of its numerical design—the structure is built mainly around one number.

It is my hope this volume will go some way toward bringing out of the shadows one of the greatest of all jewels in the treasure chest of mathematical metaphor. It would have been wonderful to come to the end and been able to identify the poet, but I remain unsure he left enough information for us to do that. He did, however, leave behind a 'mathematical signature' which may enable us to eventually discover some more of his poems.

The *Pearl* manuscript is a work of glory, of sublime and simple mathematics woven together into a cloth of gold. It is my belief it is part of a long-standing tradition which fuses mathematics and literature and is not only worth studying, but worth reviving.

Anne Hamilton
Seventeen Mile Rocks, 2013

Glossary of Mathematical Terms

Aliquot factor –
an **integer** that divides exactly into a number

Arc length –
the length of an arc (part of the **circumference** of a circle)

Circumference –
the length of the line around the outside of a circle

Decimal fraction –
a **fraction** which uses the decimal system and the place values associated with it

Decimal system –
a counting system based on the number ten in which the position of the numbers determines their value: for example, the value of 5 in 500 or 5000 or 5000000 is vastly different. In the Roman system, C always stood for 100, wherever it might have been placed.

Denominator –
the number in the bottom line of a **fraction**

Diameter –
a straight line across a circle passing through the centre and touching the periphery at both ends

Fibonacci numbers –

numbers found by starting with 0 and 1 (or 1 and 1) and adding together the last two numbers in the sequence to obtain the next one: 0, 1, 1, 2, 3, 5, 8, 13, 21, 34, 55, 89, 144, 233, 377...

Fractions –

a 'broken number', a number expressing part of a whole, an expression showing the quotient of two quantities (that is, **numerator** divided by **denominator**)

Golden ratio –

A concept going back in the history of mathematics to the time of Euclid where it is first explicitly mentioned in the following: 'A straight line is said to have been *cut in extreme and mean ratio* when, as the whole line is to the greater segment, so is the greater to the less.' Medieval scholars rightly observed this could happen at two points on a line (approximately 0.382 and 0.618 of way along the line), although this fact is generally overlooked today. 0.618 has been popularised as the golden ratio in such books as *The Da Vinci Code*, however it is merely an approximation. The related term *golden section* (in German, *goldener Schnitt* or *der goldene Schnitt*) is first known have been used by Martin Ohm in the 1835 second edition of his textbook *Die Reine Elementar-Mathematik*. The first known appearance of this term in English is in James Sulley's 1875 article on aesthetics in the ninth edition of the *Encyclopedia Britannica*. See also **phi** and **tau**.

Integer –

positive or negative whole number

Numerator –

the number in the top line of a **fraction**

Perfect numbers –

numbers whose **aliquot factors**, excluding the number itself, add up to the original number: the first seven are 6, 28, 496, 8128, 33550336, 8589869056 and 137438691328

Perimeter –

the length around the outside of a geometrical figure

Phi –

Φ (big phi) – a growth factor of approximately 1.618 commonly confused with golden ratio

φ (little phi) – a fraction of approximately 0.618 which, when set in proportion to 1, is the golden ratio

Advocated by William Schooling after a suggestion by Mark Barr for Phidias, the designer of the Pantheon, and popularised by Theodore Cook. Used predominately outside Europe. See also **tau**.

Place value –

a notation in which the position of the number determines its value (eg 73 = 7 × 10 + 3 × 1 while 37 = 3 × 10 + 7 × 1)

Prime numbers –

numbers that have no factors other than themselves and 1

Radian –

a unit of angular measure. One radian is the angle subtended at the centre of a circle by an arc equal in length to the radius. This, in degrees, is approximately 57°17'44.6" —a double radian, dubbed in this text 'dourad', is twice that and corresponds to the same length as a diameter.

Reciprocals –

a fraction where the **numerator** and **denominator** are reversed; a 'tipped over' fraction

Roman Numerals –

a system that does not use place or position to determine the value of the number. As a consequence, it has no zero to use as a 'placeholder'. I is 1, V is 5, X is 10, L is 50, C is 100, D is 500, M is 1000.

Significant digits –

those figures used to indicate the precision of a number. Leading and trailing zeroes are placeholders which indicate the degree to which the number has been rounded off.

Tau –

τ —a fraction of approximately 0.618 which, when set in proportion to the unit value 1, is the golden ratio. Generally used in Europe in preference to **phi**. Said to be from Greek *tome*, to cut.

Tetrakys (Tetract) –

The fourth **triangular number**: 10 dots arranged in a triangle.

Triangular numbers –

Numbers which can be drawn in dots so that they form a triangle. They can be simply found by adding up the counting numbers. The first 20 are: 1, 3, 6, 10, 15, 21, 28, 36, 45, 55, 66, 78, 91, 105, 120, 136, 153, 171, 190, 210.

prelude

THE COTTON LIBRARY WAS a priceless collection of old manuscripts acquired by Sir Robert Cotton (1571–1631) and largely assembled after Henry VIII ordered the dissolution of the monasteries during the English Reformation. It included a number of unique items, such as *Beowulf* and the Lindisfarne Gospels. It also contained a manuscript, previously owned by Henry Saville of Bank in Yorkshire, containing four alliterative poems written in a Midland dialect and dated to the late fourteenth or early fifteenth century.

The survival of this manuscript is quite remarkable. In October 1731, it was rescued from a fire which broke out in the Westminster home where the Cotton collection was housed. Although many priceless works were destroyed, the librarian, Dr Bentley, was able to save some of the most precious.

The so–called *Pearl* manuscript, now in the British Library, is named after the first word in its first poem. It is also known as *Cotton Nero A.x*—recalling a colourful pre–Dewey decimal cataloguing system: it was the tenth manuscript on the first shelf under the bust of the Emperor Nero in the Cotton library.

It contains the only surviving copies of *Pearl*, *Purity*, *Patience* and *Sir Gawain and the Green Knight*. All the poems are anonymous. The first and last of them are widely acknowledged as the two of the finest examples of English vernacular poetry ever written. The stellar quality of the verse outstrips Langland,[19] Gower[20] and Lydgate[21] and rivals Chaucer[22] at his very best. So how could such a distinguished poet have disappeared so utterly without trace?

Some scholars have suggested the uneven quality of the poems militates against a single author. On the other hand, there are similarities which suggest just the opposite. Opinion on whether one author or several are responsible seems to ebb and flow as regularly as the tide.

It is possible the poems are bound together for no better reason than that the scribe chose to collate four he had at hand which happened to be composed in the same dialect. Simple chance brought them together. Simple chance also excluded the short pious tale of *Saint Erkenwald* written in the same dialect, a poem which has tantalised some scholars with the possibility it was written by the *Pearl* poet.

On the other hand, many other scholars contend there is no element of chance involved at all: there is one and only one writer of all four poems—the so-called *Pearl* poet or *Gawain–Pearl* poet.

None of the poems have titles. The common names by which they are known have been long-established. The first poem in the manuscript is 1212 lines long and is universally known as *Pearl*. Recognised for its smooth-flowing and seemingly effortless rhythm, the sumptuous texture of its language and its dazzling technique, it is the story of a man who falls asleep near the grave of a beloved child and dreams of her as a queen in heaven. Like the last poem in the *Cotton Nero A.x* manuscript, *Sir Gawain and the Green Knight*, it has 101 stanzas but the 'rugged muscular' rhythms of the latter contrast strongly with the dream–like, sometimes languid, serenity of *Pearl*.

The second poem in the manuscript is sometimes called *Purity* and sometimes *Cleanness*. It is 1812 lines in length and is a curious history of God's dealings with mankind, paying particular attention to his judgments on those who are sinful, filthy and unrighteous. The story of Noah, of Abraham pleading for the cities of Sodom and Gomorrah, and of Daniel interpreting the writing on the wall as Babylon is besieged by the armies of Cyrus are all featured.

The third poem is the shortest. It is 531 lines long and is known as *Patience* from its first word. It begins with a discussion of the Beatitudes but most of its length is devoted to a re-telling of the story of Jonah, including his rejection of God's call, his time in the belly of a fish and his preaching in Nineveh.

Sir Gawain and the Green Knight, the last poem in the manuscript, is the longest. 2531 lines in length, it belongs to the tradition of Arthurian romance and the tales of Camelot.

King Arthur is at table on New Year's Day with his court, waiting for a marvel or a miracle to occur so that the feasting can begin. In rides a knight on a horse, both wholly green, an axe in one hand, a holly–branch of peace in the other. This half–giant claims the right to a game—if someone will cut off his head, he reserves the right to return the blow in a twelvemonth. Gawain takes up the challenge and beheads the knight. The giant picks his head up and insists Gawain keep the appointment at the Green Chapel, a year hence.

Towards the end of the year, Gawain sets off. The Arming Sequence which describes his gorgeous apparel and goes on to detail, at considerable length, the significance of his heraldic device—the *trawthe-laden pentangle*—is one of the most famous sections of the poem.

In desperate straits, in the dead of winter, Gawain arrives at the castle of Hautdesert and is welcomed by the host and his lady. Feted and feasted over Christmastide, he is informed the Green Chapel he seeks is only a short distance away. The host suggests he rest up the last few days and persuades him to take part in a game. Whatever the hosts wins during his hunting expeditions he'll exchange for whatever Gawain can win during his days of rest.

It's a trap. The host's wife tries to seduce Gawain but gets no more than a set of kisses. In accord with the exchange–of–winnings covenant, Gawain archly bestows them on the host. However when he is offered a magic girdle to ensure his safety during the ordeal at the Green Chapel, the temptation proves too much.

Gawain rides off with the magic girdle, dreading the ultimate test of his honour and little realising it is not ahead, but behind him. It has already taken place. What appears to have been a literary diversion turns out to be nothing of the kind. The witty dalliance between Gawain and the host's wife which seems to be in line with rambling continental romances is not a simple courtly interlude. Neither is it a natural fall in the tension before a sharp rise to the main crisis of the narrative. It *is* the climax.

This clever twist in story-telling technique lends an air of freshness to the story. This is a tale in which nothing is quite as it seems. And this is an important consideration when it comes to its mathematical design, as well.

It should be noted that this broad outline and general summary hardly does justice to the poem's humour, let alone its delicate and subtle touches, its interlaced contrasts and, above all, its exuberant joie de vivre. Silverstein calls it a 'high comedy'[23] and, despite the inescapable and deep theological overtones evident in the mathematics, it's tempting to agree.[24]

In every way, it is a counterpoint to the three poems which precede it. It dips into Arthurian legend while the remainder are drawn from the wellspring of Scripture and emphasise various well-known Bible stories or parables.

Yet each of them is so very different technically that, as Robert Blanch and Julian Wasserman have pointed out, critics who actually do consider all four poems to be the work of one writer almost always pay only lip service to that notion and treat them as isolated entities.

Admittedly it is impossible to be sure there is one author, but it is my view the *Pearl* manuscript has a single vision. I intend to demonstrate an overriding unity to be found in the theme of the Four Daughters of God—Peace, Righteousness, Mercy and Truth. The mathematics of the manuscript corroborates this theme and reveals not just a coherence of purpose but a co-inherence—a realm of meaning—which cannot be accidental.[25]

The four poems clearly have a single architect—one and only one mathematical designer. While each poem exists as a separate piece, none of them make complete mathematical sense until they are fitted together like a puzzle box. Each different combination of parts within the puzzle is like a module with a unique sense of its own, governed by its own mathematical 'token' but also by overarching numerical themes. Everything is linked and locked as the poet tells us on Line 35 of *Sir Gawain and the Green Knight*—'with lel letteres loken'—so that if there is one pathway to a significant arithmetic crossroads, there is at least another as well.

Since it is reasonable to assume the mathematical architect and the poet were one and the same person, I conclude there must have been a single author. He was deeply religious, highly intelligent, exceptionally perceptive and so numerically creative it is difficult to find a jury of his peers to judge his talent.[26]

JJ Anderson observes that the poet is capable of pouring himself into very different imaginative moulds and that there may be 'an element of self–challenge about this; there are hints in the metrical and numerological (*sic*) constraints that he works with that he is interested in setting himself a difficult task and then seeing how well he carries it out.'[27]

While I agree, there's far more to it than simply the satisfaction of achieving a self–imposed goal. Ed Condren rightly describes Chaucer's *Parliament of Fowls* as a mathematical *tour de force.* Yet I feel it is still reaching skywards for the fistful of stars the *Pearl* poet already possesses. Perhaps a numerical literary undertaking as intricate and comprehensive as the *Pearl* manuscript will eventually be found but I would not be at all surprised if it really was the zenith of the style.

In 1956 Ormerod Greenwood proposed Hugh de Masci as the identity of the *Pearl* poet. Greenwood's guess has at times been widely accepted, at other times vigorously disputed. The consensus has varied continually as different evidence has emerged. Greenwood originally proposed Hugh de Masci as the author because

(a) The Cotton Nero *A.x* manuscript had 'Hugh de' inscribed at the front

(b) Masci or Massey was an old name in Cheshire where he contended the manuscript was composed

(c) The number of stanzas common to both *Pearl* and *Sir Gawain and the Green Knight*—101—is CI when written in Roman numerals. He suggested de Masci had encoded the last syllable of his name using fairly standard numerological practice.[28]

A few decades later, Clifford Peterson took Greenwood's idea further, suggesting the surname was right but Hugh should be amended. He proposed that I. Massi or J. Massey was the mysterious *Pearl* poet.

J. Macy was detected as an encoding amongst the ornamental designs of *Purity*. Furthermore, the anagram I. de Masse appears to be encoded in *St Erkenwald*[29] and, according to Katherine Adam[30], John Massi may be contained as an acrostic in the 76th stanza[31] of *Pearl*.

Ingenious as the CI solution for 101 is, I don't think the answer is primarily numerological. Much as I agree that John Massey, rector at Ashton–on–Mersey at the turn of the fifteenth century, is the most logical candidate[32] advanced so far in the authorship debate, I doubt the number is a cryptogram for any part of his name. If 101 does serve a dual purpose, any connection to the writer himself is at best an afterthought. Nothing more than a case of a happy accident.

Because, quite simply, 101 is a mathematical metaphor that points heavenwards.

Chapter 1

Angels singing: the number 101

Music is well said to be the speech of angels; in fact, nothing among the utterances allowed to man is felt to be so divine. It brings us near to the infinite.

Thomas Carlyle

101 IS ARGUABLY THE MOST obvious number associated with the *Pearl* manuscript. Arguably, because 5 and 12 are so frequently mentioned in association with *Sir Gawain and the Green Knight* and *Pearl* respectively. However no mystery clings to them.

101 on the other hand is the number of stanzas in both *Pearl* and *Sir Gawain and the Green Knight*, the first and the last poems in the manuscript, and has puzzled readers for over a century. Is it simply coincidence? Is it a numerological encoding, as Greenwood and others have suggested, for part of the author's name?

If not, is it nonetheless an indication that the author of *Pearl* and *Sir Gawain and the Green Knight* are the same person? Many scholars think the repetition of 101 supports this case and the prevalence of comments about it are generally along those lines.

Is it a clue and, if so, to what? If it's not a clue, is a mistake?[33]

At first sight, 101 is a perplexing number since it fails so obviously to be 'perfect', not by falling short of 100 but by overshooting the mark. In fact, it indicates to some reviewers that the poet deliberately chose to 'flaw' his work to acknowledge that only God creates perfection.[34]

This view, however, is not the only one. It is countered by the suggestion that perhaps 101 is actually the equivalent of 100 in a similar sense that the old 'year and a day' and the modern year[35] are the same thing. As eight days were once considered to constitute a week.

In mathematics, we might consider this as a fencepost problem: there are 101 posts needed to position one post every metre along a hundred metre fenceline. In this instance, it is about the nature of counting and whether it is inclusive or exclusive. Inclusive counting includes both endpoints while exclusive counting does not.

101 has some interesting attributes: it is the 26th prime number[36] and the 13th partition number.[37] Neither of these seems particularly relevant, however, to the poet's motivation in choosing it. In fact, all such solutions, even the highly likely one that counting begins *after* 1 which as the sacred monad is a symbol of God,[38] smack of special pleading.

It should be noted that these two occurrences are only the tip of the iceberg. There are several obvious multiples of 101 in the design of the manuscript.

- In Fitt III of *Sir Gawain and the Green Knight,* there are 303 lines (3 × 101) describing Day 3 of Gawain's stay at Hautdesert.
- Two of the poems—*Sir Gawain and the Green Knight* and *Purity*—together total 4343 lines (43 × 101).
- The combined total of the three poems, *Sir Gawain and the Green Knight*, *Pearl* and *Purity*, equals 5555 lines (55 × 101).
- Kent Hieatt pointed out that the long lines of *Sir Gawain and the Green Knight* end on Line 2525 of that poem. He used this fact as an important example of its 'fives'—however, they also demonstrate the significance of 101 since 2525 is 25 × 101.

There are many other examples of multiples of 101 which will crop up as we analyse the design,[39] but even these few serve to illustrate the point that 101 is even more integral to the text than is commonly supposed.

I would also like to point out that 101 is a numerical motif in a very significant text, particularly when we are considering *Sir Gawain and the Green Knight.* Paul's letter to the Ephesians opens in Greek with a massive 202–word sentence, so complex in its structure that there is more than one possible interpretation. The letter finishes up with the famous passage detailing the Armour of God, often referred to as the inspiration for Gawain's panoply in *Sir Gawain and the Green Knight.* This closing section of Paul's epistle has 101 words in Greek.[40]

So because there is a theological precedent for 101, it is quite disappointing that the manuscript as a whole does not feature a multiple of this number. At 6086 lines,[41] it is some 26 lines longer than 60 × 101. It almost seems that an emerging pattern has broken down even before it has fully surfaced. Instead of being a multiple of 101 at 6060, it is close to 101.43, a percentage difference so trivial that, if I were to apply the usual rules of scientific testing, would not even warrant an explanation.

However, I intend to take the attitude throughout this book that it is important never to despise small things. Indeed, even at this point, any serious student of music theory may already have become suspicious about 101.43 and guessed the possible allusion behind the extra lines. They are not insignificant: they actively offer the best possible clue toward unveiling the 101 metaphor.

However before tackling the question of what 101 signifies and why the poet chose it, I am going to take a leaf from his repertoire and make a short digression.

I have a question about his mathematical inspiration: *Is 101 the number around which the manuscript as a whole is organised?*

The short answer is *yes*—fundamentally it is.

My next question is more subtle: *Is it also the defining number of the manuscript?* That is, is it the mathematical metaphor at the heart of the text? Is it the numerical token which encapsulates the poet's religious and philosophical vision and expresses the conceptual unity of his thought?

No.

I don't believe it is.

Yet if it is not his inspiration, there must be a well–spring hidden behind it.

So where does it come from? What is the defining number of the manuscript?

Ed Condren in his analysis of the *Pearl* manuscript suggested that the poet crafted each poem individually using mathematical motifs which were inspired by the intricate carpet pages of illuminated manuscripts such as the Book of Kells or the Lindisfarne[42] gospels. For *Pearl*, he suggested that the poet used the mathematics of a dodecahedron (the solid figure with twelve faces which Plato regarded as emblematic of the shape of the universe[43]). This would account for its 'twelves' and for some of its internal features.

His analysis concentrates mainly on uses of the 'golden ratio' to inter–connect the text. He states that the organising principle of Cotton Nero *A.x* is the golden section.[44]

When it came to the irregular stanza construction of *Sir Gawain and the Green Knight,* he proposed an even more complex template. Taking a hint from the unusual heraldic device painted on Gawain's shield—the five–pointed star described in Lines 619–669—Condren suggested that the poet organised his work around the results obtained from drawing a figure of a cross overlaid on a pentagram surrounded by a double circle.

A cross.

Overlaid on a pentagram.

Surrounded by a double circle.

I have to admit that such an elaborate motif is so mathematically daunting, it would be an understatement to say I had considerable doubts about it. Condren's suggestion that the design was taken from a geometric drawing, rather than arithmetic, hardly felt satisfactory. It was difficult to imagine a poet manipulating his line count just to fulfil the constraints of such an intricate and complex design—at least for the sheer love of mathematics alone. Unless there was a better reason, it didn't seem even remotely likely.

When Condren's analysis failed at a critical juncture to draw out an unequivocal and unique mathematical solution to the size of the cross allegedly underlying the design, it was almost enough to dismiss his theory.

Almost.

One of the oldest methods of solving equations is 'guess and check'. Condren guessed but did not check. Thus he missed some of the wider implications of his answer.

I admit I had the utmost reservation about his entire analysis until the very last page. I could believe a poet would embark on a mathematical challenge but not to the degree Condren suggested.

Such a poet would have to think mathematically. A true mathematician has a certain psychology: delighting in clarity and elegance, not in an impenetrable thicket of tedious calculation. And despite the difference of six centuries, I really can't see this changing much. It's a truism that the more mathematically–inclined an individual is, the less mathematics he is likely to use to solve any problem.[45] It's about strategy, rather than slog.

In addition, there was another and more important reason I felt caution: both the dodecahedron and the pentagram that Condren suggested was underlying the work required the use of the golden ratio which, in the fourteenth century, should have been approximated by a fraction. However any common fraction, repeatedly applied, would create an increasing margin of error. But the error was not consistent. Condren could not discover either the fraction used or the method used to minimise the error.

The thing was—neither could I. That troubled me; it shouldn't have been a difficult problem to solve. The messier it seemed, the less believable it got.[46]

The fact is, as Stephen Hawking has admitted, the biggest blunders of understanding are usually made not because an analysis is too complicated. They happen rather because the call of simplicity is as seductive as the song of the sirens. The power of beauty has enormous sway in physics and mathematics. It is held, almost as an article of faith by many mathematicians, that the right answer is exquisite.

Now by proposing that *Pearl* was based on the mathematics of dodecahedron, Condren was automatically theorising that the poet was a numerical artist of stunning ability. Compounding this suggestion with the idea that a cross, a pentagram and a double circle underpinned *Sir Gawain and the Green Knight*—even one based around geometry rather than arithmetic—catapulted the poet into an even more rarefied class.

The only obvious similar design to this combination of pentagram, cross and circle is Leonardo da Vinci's *Vitruvian Man.* Was the *Pearl* poet a poetic and mathematical genius as wide–ranging in his interests as Leonardo? Certainly, he was good at what he did, but was he *that* good?

As we shall eventually see, some astonishing similar mathematics underlies these different artforms.[47] It is a mistake, however, to think that this means the poet's work should be more complicated. I still hold the view that it should be less: a gifted mathematician works as simply as possible.

So until Condren revealed the guess he believed defined the size of the cross, I was unconvinced. The sheer complexity of what he proposed seemed to militate against the design. When he did unveil the final number, however, there was a seismic shift in my perception of the whole manuscript. His answer, although not formally derived, was so convincing it was instantly clear he had unearthed the defining number of the text.

The right answer, after all, must be exquisite. It must be mathematically elegant and it must fit perfectly in every way. Ultimately, it must be obvious.

And this answer was.

It had been staring every reader in the face all along: 490, the number of lines in the Fitt I of *Sir Gawain and the Green Knight.* An arithmetic metaphor par excellence, the meaning of 490 would have needed no translation in the fourteenth century.

Then Peter came to Jesus and asked, "Lord, when my fellow believer sins against me, how many times must I forgive him? Should I forgive him as many as seven times?"

Jesus answered, "I tell you, you must forgive him more than seven times. You must forgive him even if he wrongs you seventy times seven."

Matthew 18:21–22 NCV

490 is a mathematical token for forgiveness, the quality of mercy expressed quantitatively, a numeric metaphor for reconciliation.

In fact, turning to *Purity*, the second poem in the manuscript, we find that Line 490 does indeed specifically refer to reconciliation.

Moreover, there is a natural association in Christian theology of forgiveness and reconciliation with the Cross of Christ. This lends credence to Condren's suggestion that the design incorporates a cross with a 'crossbeam' of 490 lines.

On the other hand, it's fair to say this may be reading too much into the size of a single Fitt. If there isn't any real mathematical support for this guess—unless either the alleged pentagram or the double circle leads swiftly to other prominent numbers in the design—then this is an ingenious answer but not an irrefutable one.

Of course if it led to 101, that would be a very different matter.

There are, in fact, two possible ways to derive 101 from 490 which rapidly bring to light other significant mathematical features of the text. Condren did not examine either of the following methods, but his intuitive surety with the underlying design immediately sheds light on both the alternatives I am about to outline. As the trail of arithmetic evidence favours both methods equally, it is impossible in my view to choose between them.

The first method uses the golden ratio (which every pentagram famously possesses) and the number three. The second method uses the symmetry of a cross and the square of the number 12.

Method A
(based on the mathematics of a pentagram)
490 × τ = 303
303 ÷ 3 = **101**

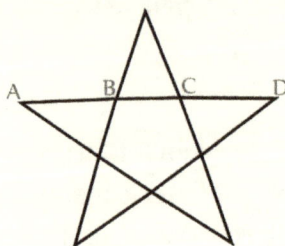

If length of AD = 490,
then length of AC = 303

By using τ ≈ 0.618 for the golden ratio, Method A immediately produces 303, the number of lines within Fitt III of *Sir Gawain and the Green Knight* which correspond to Day 3 at Hautdesert. This is the very day where Gawain takes the green girdle and is in most need of forgiveness[48] for his actions. Thus it links the size of the Fitt I to the relevant section of Fitt III using the mathematics of the pentagram[49] detailed in Fitt II within the description of Gawain's shield that happens to start right after Line 618.

Take note of that number and glance back at the value of τ at the beginning of this paragraph. I find it difficult to believe it is simply random chance which lines up the description of a heraldic pentangle so it begins immediately after the line which contains the significant digits[50] of the golden ratio. That's too coincidental entirely.

Particularly when this method is illustrated by a beautiful transition within the text, carefully signposted all the way.

Method B
(based on the mathematics of a crucifix)
490 ÷ 2 = 245
245 – 144 = **101**

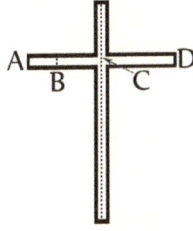

If the length of AD = 490,
then length of AC = 245,
where AB = 101 & BC = 144

Method B on the other hand recognises that the crossbeam of a crucifix is cut in half by the upright. Each half of the horizontal bar could then be further partitioned in an enormous number of different ways, but for symbolic purposes, what better break–up than 144 and 101?

144 is a factor of the combined length of the first two poems (1212 + 1812 = 3024 = 21 × 144) but it appears most importantly, of course, in *Pearl* as the significant digits in 144000, the number of virgins who are brides of the Lamb, an image harking back to the book of Revelation. It is therefore almost certainly a mathematical metaphor in its own right.

Yet what does 144 stand for? Almost certainly, I feel, some concept akin to the 'communion of saints' or the 'community of the elect'.

Method B undoubtedly has the edge for those of us who feel uncomfortable with irrational numbers, but it's clear the poet was unfazed by them: the size of *Purity* is related to 101 through the golden ratio (1812 × τ^6 ≈ 101).[51]

Whichever way 101 is viewed, it is linked back to 490 and it is my opinion the poet was making use of both methods and they were equally important to him. They serve to confirm the importance, not of 101, but of 490 as the source[52] of 101.

Now it would be all too easy to assume that 101 is simply a connector and served no other numerical purpose than to link one mathematical metaphor to another.

However, because the total length of all four poems is 6086 lines, I am far from convinced this is the case. Individually the poems are:

PEARL	1212 lines
PURITY	1812 lines
PATIENCE	531 lines
SIR GAWAIN & THE GREEN KNIGHT	2531 lines

The first thing to look for in any set of numbers is a pattern. *Pearl* and *Purity* both exceed an even hundred by 12 lines while *Patience* and *Sir Gawain and the Green Knight* exceed it by 31 lines.

Following Condren's example, I intend to ignore those 'extra' lines—Bede's 'left hand' numbers'—for the moment and re-configure:

PEARL	1200 lines
PURITY	1800 lines
PATIENCE	500 lines
SIR GAWAIN & THE GREEN KNIGHT	2500 lines

This simplified tabulation shows the total length of the first two poems is the same as the total of the last two (1200 + 1800 = 500 + 2500 = 3000). This means there is a 1:1 ratio between the two 'halves' and a 2:1 ratio between the 'full' manuscript and either 'half'. Moreover there are some other simple ratios in the table:

PURITY: *PEARL*	= 1800:1200	= 3:2
First 'HALF': *PEARL*	= 3000:1200	= 5:2
First 'HALF': *PURITY*	= 3000:1800	= 5:3
SGGK: *PATIENCE*	= 2500: 500	= 5:1
Second 'HALF': *PATIENCE*	= 3000: 500	= 6:1
Second 'HALF': *SGGK*	= 3000:2500	= 6:5

Condren, on observing these superbly clean results, suggested that a musical formulation was the guiding principle.[53]

Augustine defined music as *bene modulandi*, the 'science of measuring well',[54] so a design combination of either arithmetic and music or geometry and music would certainly not be out of the question since all three were related studies in the quadrivium.[55] Only astronomy is missing from the present line-up.

Indeed Augustine's statement that in *De Musica* that arithmetic, geometry, astronomy and music are sciences of number through which the mind is raised from the contemplation of 'changeable numbers in inferior things to unchangeable numbers in unchangeable truth itself'[56] suggests even more: the possibility that there are theological as well as musical undertones here should not be ignored.

CS Lewis, hardly the most mathematically-minded critic, nonetheless recognised that metaphors of proportionality are seeded through the New Testament not so much as dogma but as 'flavour or atmosphere.'[57]

With regard to music, simple numerical ratios (as for instance in the lengths of string or the weights of hammers) had long been known to produce the most effective harmony. In the Middle Ages, musical intervals were considered to be multiple, superparticular and superpartient.[58] Multiple intervals are represented in the *Pearl* manuscript by 2:1 (diapason or octave) and also by 5:1 and 6:1 while superparticular intervals are shown by 3:2 (the fifth or diapente) as well as 6:5.

The relationship between music and mathematics went back to antiquity. In the sixth century BC, the mathematician and philosopher, Pythagoras of Samos, was alleged to have conducted the first empirical measurements which led to the theory of music. Given this long association and also given the intimacy of mathematics and music in the quadrivium, it would perhaps be surprising if there were no musical features at all factored into the design of the *Pearl* manuscript. It does indeed seem that, once those 'extra' lines are discounted, there is a mysterious background melody playing quietly away.

However—the dismissal of those extra lines smacks of special pleading.

A poet who can design a poem around the mathematics of a dodecahedron should surely have been able to get his musical theory right. It's almost as if he were deliberately and consciously avoiding multiples of 10 or 100—he's got an extra 1 over and above the 100 neat stanzas, an extra 12, an extra 31.

Actually he takes us beyond theory and out into the real world. When it comes to genuine musical instruments, the ratios aren't quite so perfect as those whole numbers suggest. An octave isn't produced by using one string exactly twice the length of the other. A tiny discrepancy known as the diatonic or Pythagorean comma upsets the neat numerical proportions. It was for this reason that well-tempered scales took centuries to produce.

The mathematical property of a diatonic comma can be expressed as slightly more than 101% and so it was seen as an extension beyond the whole, as a stretch—on tiptoe admittedly—toward infinity. Thus a diatonic comma was a symbol of the harmony of heaven, the sweet song of the angel host and the celestial Music of the Spheres.

101 as an allusion to a diatonic comma is therefore a mathematical metaphor combining both music and astronomy.

Taking the *Pearl* manuscript be to without excess (that is, 6000 lines in length), let us consider it as a 'musical instrument' which encompasses the heavenly Music of the Spheres.[59]

Let us now therefore calculate the exact size of a diatonic comma for a 'musical instrument' of 6000 lines. A diatonic comma is not exactly 101%, so the total length of the manuscript should emerge as 6060 lines only if the poet were 'rounding off' his calculations. A more precise answer is 6082 lines.[60]

This is so close to the actual length of 6086 lines that it seems almost pedantic to try to explain the discrepancy. Nevertheless, lest it be seen as special pleading to ignore these four lines, let us note the three possibilities for this miniscule error.

(1) the poet knew the precise mathematical size of the comma (which is $3^{12}/2^{19}$) but he made a small mistake in calculation

(2) the poet did not know the size of the comma to our current accuracy but he did know it with sufficient precision to achieve a result within 4 lines

(3) the poet knew the precise mathematical size of the comma but deliberately chose to add 4 lines

If I have to choose between the scenarios above, I believe the third option is most likely to be the correct one. There's no accident in those four lines—they were deliberately added. I believe there is particular mathematical reason for this: 6086 can be evenly divided by 17.

Many reviewers, critics and scholars of today, having not put aside a romantic nineteenth century vision of the Middle Ages tend to see the *Pearl* poet as a wordsmith who welded numerology, Celtic nature religion and ancient echoes of magic to the traditional rituals of medieval Christianity. Anyone who thinks that way had better read the last seven words of the previous paragraph again: *6086 can be evenly divided by 17.*

The significance of this may not be immediately apparent. However the prevalence of 17 within the deep fine structure of the text suggests this line count is the poet's final declaration of his ultimate intent.

In case anyone should have missed, or even overlooked, this final avowal, he deliberately wrote the last line in Old French. Just to make sure everyone got the point.

It wasn't syncretism. Exactly the opposite.

It was war.

digression 1

The *Pearl* poet was far too meticulous to have made a mistake of four lines in setting up a diatonic comma. As we progress, we shall see that there are many other mathematical constraints which affect the length and which create tiny discrepancies such as this one. The poet was striving to link one arithmetic token to the next using only a few mathematical ideas with theological overtones. Sometimes this works brilliantly while at other times it requires considerable ingenuity.

In favour of the view that 17 was the dominant consideration for the overall length is the fact that the stanzas describing the pentangle in *Sir Gawain and the Green Knight* are a total of 51 lines long—and therefore also a multiple of 17.[61]

There are many more instances which will be noted as we progress.

It is towards the end of *Sir Gawain and the Green Knight* that Sir Bertilak of Hautdesert suddenly reveals that the Beheading Game was instigated by Morgan 'the goddess'. Is this a sop towards supplying the characters of Bertilak and his wife with motivation as some commentators suggest? Or is it a clue, not to the plot, but to the design?

By 'goddess', I think the poet intended to direct the reader's attention to the sisterhood of priestesses in the history of Geoffrey of Monmouth rather than to the witch–queen of the later romances. It is worth remembering in this context that Geoffrey claimed Morgan was famed for teaching mathematics.[62]

Indeed, it is quite possible that, far from being the disappointing, throwaway gesture that modern critics of the poem find it, Morgan's unveiling was intended to be deeply satisfying to any medieval reader who had spotted the mathematical clues along the way. It would have confirmed suspicion, not been a sudden inexplicable 'deus ex machina'.

In fact, the placement of Morgan's name is 6001 lines into the text. Not only is this divisible by 17, the section which explains her role in the Beheading Game covers an additional 17 lines.[63]

So I make no concession to the view that the recurrence of 17 is simple coincidence and that a poet has better things to do than count lines.

Stichometry—line counting—already had a long and respectable history going back to the halcyon days of Plato and Greek poetry. The Hebrew *Sopherim* of the fourth century were said to be so concerned with getting the count right with regard to each jot and tittle of the Torah that they argued whether the middle letter and middle verse belonged to its first or second half.[64]

The *Sopherim* are well-known from the New Testament as the 'scribes'. Generally thought of today as writers, their name actually means *counters*.[65] A scribe's three main tasks were:

(1) copy the Scriptures verbatim

(2) set the Scriptures in correct order

(3) count each letter, line, syllable, jot and tittle to ensure the text was duplicated exactly, word for word, so that not even the smallest missed stroke of the pen would happen or an element thereby drop out of the Law.

The job of the *Sopherim* was to cross-check copied manuscripts and to testify to the truth of their transmission. Their mathematical 'proof-reading' provided independent verification of the copyist's work. It wasn't a simple matter of adding: the mathematical design often included so-called square, rectangular, triangular and perfect numbers incorporated into the text.

As David Howlett points out regarding later examples of this type of design—he terms it 'Biblical Style'—the mathematical design 'self–authenticates' the text. The more complicated the design, the harder it is to tamper with, either deliberately or accidentally. In the days before photocopies and scanners, it was crucial to have internal verification of the message.

If, for instance, Patrick sent out a pastoral epistle denouncing one of the local kings which was to be disseminated by being copied and read in various Celtic churches, it was vital for the recipients to have some way of knowing they'd received the genuine words of the saint. It may well have been a matter of life and death.

In fact, as early as the first century, the authentication of pastoral letters was already a crucial matter. In his second letter to the Thessalonians, Paul mentions a forged letter purporting to have come from him to the effect that the 'day of the Lord' had arrived.[66]

Internal mathematical elements would not necessarily have been a safeguard against fraudulent missives but they would have been against corruption of the text.

Going back to the early gospels and epistles, this 'self–authentication' combined with an element important to Greek literature: the idea of textual beauty in terms of its mathematical design. Maarten Menken mentions Isocrates who apologises to the reader, in one instance, for going outside the ideal canons of beauty to finish his story and, in another instance, for not finishing the story because of the need to conform to the canons.[67]

In this respect, the apostle John seems to have grasped the importance of appealing to the sensibilities of a Greek reader, while never departing too far from his Hebrew roots. Richard Bauckham reports the well–known curiosity that John's gospel starts with a 496–syllable poem and ends with a post–resurrection epilogue of 496 words. To the Greeks, with their penchant for both triangular and perfect numbers, 496 was evocative of divine beauty. To the Jews, however, the 'book–ending' of the gospel with 496 would have hearkened back to the idea of Hebrew poetry with its distinctive form: the mirror–like chiasmus. Specifically it might have reminded them of the prophecy of Immanuel within the Book of Isaiah which was 'book–ended' by 496 syllables of verse.**⁶⁸**

Now John was not idly nicknamed by Jesus as one of the sons of thunder. He might have started his gospel with one of the archetypes of mathematical beauty—the *logos*—but, like the *Pearl* poet, he lobs a bombshell to start with and finishes off with another. His first sentence about the *logos* is 17 words. It's an iconoclastic mix which would have horrified the Greek literati of the time: no one back then with a taste for 'the good, the true and the beautiful' would have put 17 and the *logos* together.

In fact—while it's hardly obvious to us at a remove of twenty centuries, the overt mention of 153, the 17th triangular number, at the end is like lobbing a grenade into the finale:

Jesus said to them, 'Bring some of the fish you have just caught.' Simon Peter climbed aboard and dragged the net ashore. It was full of large fish, 153, but even with so many the net was not torn. Jesus said to them, 'Come and have breakfast.' None of the disciples dared ask him, 'Who are you?' They knew it was the Lord. Jesus came, took the bread and gave it to them, and did the same with the fish. This was now the third time Jesus appeared to his disciples after he was raised from the dead.

John 21:10–14 NIV

153 was associated since at least the time of Archimedes with the square root of three and famously called 'the number of the fish' (from the geometric diagram describing $\sqrt{3}$). John plays with words, numbers and alphanumerics—or gematria—in this sequence.

With gematria, we are on the edge of numerology.

However it's worth remembering the words of Ronald Youngblood: *'Flagrant abuse of various forms of numerology, including especially gematria, should not be permitted to blind us to the undoubted use of numbers in a figurative sense or of numbers as a literary device in the Bible (as well as elsewhere in the ancient world.)'*

John's work might have elements of gematria as might the *Pearl* poet's. The use of 17 (and, as we shall see, 153) in both cases, however, indicates clearly the battle stance both have adopted. Their work is definitely and defiantly Christian in opposition to Pythagorean or Platonic Gnosticism. As such, it gives clear notice of their intent.

So, as far as I am concerned, chance plays no part in the fact that the *Pearl* manuscript, at 6086 lines, has a factor of 17. However, before examining his use of multiples of 17 further and how it divides him from his contemporaries, let us first consider what he has in common with them.

Chapter 2

Full Circle: the number 22

For architecture, among all the arts, is the one that most boldly
tries to reproduce in its rhythm the order of the universe, which the
ancients called 'kosmos,' that is to say ornate, since it is like a great
animal on whom there shines the perfection and proportion of all its
members. And praised be our Creator, who has decreed all things, in
their number, weight and measure.

The Name of the Rose, Umberto Eco

There is no easy way to stumble across the fact that 6086 is a multiple of 17. It's a matter of dividing and checking for a remainder. Likewise, there's no especially easy way of realising that *Pearl* and *Purity* combined are a multiple of 7 and 144—very suggestive numbers in such a context.

On the other hand, one factor that is particularly easy to recognise anywhere in any mathematical context is 11 or 22. The prevalence of 22 in medieval poetry is so striking that it is clearly a metaphor of some significance.

By the time of Milton, mathematical metaphor seems to have faded entirely: 'this little known tradition, not obvious, nor obtrusive, but retired.'[69] The custom of creating poetry which featured multiples of 11 or 22 lines seems to have lasted longest, however, since it is apparent into the fifteenth century, long after the golden section, for instance, appears to have disappeared as a common construction device.

As far as I am aware, no one has previously noticed how common 22 is. It may not be ubiquitous but it occurs far too often for coincidence. Michael Robertson noted the prevalence of 11 in *Sir Gawain and the Green Knight*[70] and Eleanor Bulatkin noted the use of 66 in the Oxford *Roland*.[71] Both these examples are in my view simple variations on the 22 motif; in fact Bulatkin acknowledges that the pattern she found may well be a combination of 44 and 22.

I first noticed 22, not in the *Pearl* manuscript, but in *St. Erkenwald*, that alliterative poem which is in the same Midland dialect as *Pearl*, *Purity*, *Patience* and *Sir Gawain and the Green Knight*. Although it was once thought that there might have been a single author for these five poems, statistical analysis currently seems to have skewed the consensus of opinion towards the view that *St. Erkenwald* was written by a different poet.

Now I have taught statistics from time to time. The very idea of applying statistics to an anonymous creative work is, in my view, a profound and appalling failure to understand the limitations[72] of the method. If any mathematical analysis is to be brought to bear on poetry, it should be the most obvious, not the most obfuscating.

This is why I have chosen to look at basic numerical design which does not require a degree in mathematics to check. It does not provide 'proof' but it will provide a body of evidence. In looking for metaphorical significance in the important numbers of the actual text and detecting patterns of mathematical thought, we will be venturing on much more secure and stable grounds than statistics.

Turning therefore to *St. Erkenwald*, we should note the total length is 352 lines, an exact multiple of 22.

Sir Gawain and the Green Knight fails to be a multiple of 22 by just one line. However, if the disputed last line in Old French (*Hony Soyt Qui mal Pence*)—often said to be added by a different hand—is discounted for the moment, then at 2530 lines, *Sir Gawain and the Green Knight* also pans out as a multiple of 22.

Whether writing off that last line is legitimate or not is an open question. Certainly, for the sake of the entire manuscript as a multiple of 17, it is a vital inclusion. Also, for the sake of the 'excess pattern'—12 over the hundred in both of the first two poems of the manuscript and 31 over the hundred in the last two poems—it is essential as well.

So, for the moment let us simply flag this as a potential anomaly and forge on. The poet, as we shall see in the next chapter, enjoyed games of arithmetic and played with numbers confidently. He was not inflexible in the way he conceived of total line length. Perhaps there were times when the poet included this last line in his numbering and others when he considered it to be dispensable.[73]

What the line does tell us, however—assuming that it has been there since the beginning and has not been added later—is that the metaphor of 17 was far more important to the poet than the metaphor of 22.

The author of *Death and Liffe*[74] similarly held the token of 17 to be his most important constraint, while the writer of *Sir Orfeo* (which fails to be a multiple of 11 by just one line) clearly felt that sacrifices needed to be made in order to establish a poetic architecture[75] built on a hidden 17.

Most writers, however, stick with simple principles and concentrate on multiples of 22 or 11. Poems featuring such multiples include:

- *Of a Rose*[76]
- *Lovely Tear from Lovely Eye*[77]
- *Man and Woman Look on Me*[78]
- *Now that Man is Hale and Whole*[79]
- *Holy Lady Mary, Mother, Maid*[80]
- *This Middle Earth was Made for Man*[81]
- *A Beauty White as Whale's Bone*[82]
- *Mother, Stand Firm Beneath the Rood*[83]
- *Divine Love*[84]
- *All Too Late*[85]
- *Nun Priest's Tale*[86]
- *Alliterative Morte Arthure*[87]

- *Winner and Waster*[88]

and, when its Prologue is excluded, *Parlement of the Three Ages*. Others, like *Summer Sunday*[89] (also called *Fortune*), are just one line out.

There are undoubtedly many more[90] since I haven't particularly set out to look for them but have simply surveyed two general collections randomly picked from my bookshelf for this list.

Cligés by Chrétien de Troyes is 6767 lines long, a multiple of 101, but if the opening of 90 lines is excluded, the remainder turns out to be a multiple of 11.[91]

While on the subject of Chrétien, the Guiot manuscript of *Erec et Enide* is 6880 lines and thus not a multiple of 22. However Joan Helm points out that each page is set out as three columns of 44 lines, in her view facilitating the author's ability to keep track of the number of lines, the most important of which were 176 (or 8 × 22) and 1760 (or 80 × 22). Similar to the use of 66 (3 × 22) in the Oxford *Roland* is the use of 11 in *Beowulf*. The three fights with the monsters begin in Fitts 11, 22 and 33.

David Howlett points out that the Gnomic poems of the Exeter Book are set in 3 collections of 6 different verse types which total 6666 letters. Not only a multiple of 22, it also has the heavenly choir number 101 as a factor.

Turning our attention for a moment from the poetry of the fourteenth century to the king's court, it is notable that even the crown of Anne of Bohemia, wife of Richard II, features this circularity based on 22. Her exquisitely crafted wreath, made of gold and extravagantly adorned with gems, eventually became part of the dowry of Blanche, daughter of Henry IV, for her marriage to Prince Ludwig III of Bavaria in 1401. The crown is currently on display in a Munich museum: *'high–pinnacled, flower–like in its motifs, and encrusted with precious jewels— just as it is described in* Pearl. *It consists of an elaborate twelve–part circlet from which rise twelve golden lilies with trefoil leaves at the tops and sides, decorated with a profusion of 132 pearls.'*[92]

132 pearls: not only a multiple of 22, but of 12 as well, the most obvious mathematical motif in *Pearl*. 132 is also, however, the

number of *regular* stanzas in *Patience* as well as the number of lines in *Summer Sunday*, a poem which at times has been considered as a poignant lament for the death of Richard, Anne's husband.

Spotting multiples of 22 is not hard. The procedure for deciding whether a poem has a factor of 22 is very simple. Consider *Sir Gawain and the Green Knight* at 2350 lines.

2530

Add up the pairs of alternate digits indicated:

2 + 3 & 5 + 0

Both answers are the same and the number 2350 is even, so this number is divisible by 22.

Let's do a similar thing for *St. Erkenwald* at 352 lines.

352

3 + 2 & 5

Again, both answers are equal and 352 is even, so *St. Erkenwald* is divisible by 22. If the answer is the same but the length of the poem is *not* even, then a multiple of 11 is in operation. If the answers for the sum of the alternate digits are *not* the same, then the poem is *not* a multiple of 22.

Simple as that.

Well, almost. There is another occasional slight complication to consider which does not concern us here.[93]

The question raised by these different occurrences of 22 is: what does it mean? Is it an arithmetic metaphor? Is it a numerological symbol? Is it neither?

Eleanor Bulatkin could be right in suggesting that the 66 pattern in the Oxford 'Roland' is a numerological construction of 6 × 11 ('the perfect number' × 'the number of excess') but what if it is 3 × 22 instead? How could that be interpreted?

If we went back to music, could we find meaning there? Certainly 22 might be a reference to a hexatonic scale. Maybe, but I doubt it.

On the other hand, 22 could be an allusion to the number of cards in a tarot pack. Probably not, given the even more obvious presence of 17.

Of course, it might be just as easily refer to the number of letters in the Hebrew alphabet.

It may simply be mathematical; after all, it is the 8th partition number and, given that 101 is the 13th partition number, there's a remote possibility this is the reason the *Pearl* poet selected it. However, that doesn't explain anyone else's choice.

I am innately biased towards mathematics, I admit, so I prefer to explore the obvious mathematical connotation. Even into my lifetime, 22 was intimately linked to a circle.

For centuries, the ratio between the circumference of a circle and its diameter was cited as $^{22}/_7$: a fractional approximation for the irrational number π which had been known since at least the time of Archimedes.[94]

A circle may be the answer, but what purpose does it serve? Why bother unless it has some worthy meaning?

Some poems in the *Pearl* manuscript show a 'circular format' in the sense that they start and finish with the same idea; sometimes in fact the very same words.

There are five ways suggested by Matthew of Vendôme in his *Ars Versificatoria* for ending a poem[95] and also one natural[96] and three artificial ways[97] of starting a poem recommended by Geoffrey de Vinsauf in *Nova Poetria*. However none of these *quite* fit this format of circularity particularly favoured by the *Pearl* poet. This is unusual because Chaucer, for example, pays close attention to the advice of both Matthew and Geoffrey. On the other hand, Geoffrey's advice about constructing poetry has not been completely ignored: 'The work of conception, he maintained, was to be developed with 'a circular structure in mind, a common shape, often subclassified as a 'rose' or other sort of wheel, or...*mappa mundi*.'[98]

Here's an interesting thought—a 'rose' or a wheel: something circular.

Pearl begins 'Perle plesaunte, to prynces paye' and ends 'And precious perles vnto His pay', a widely recognised circularity which links Line 1212 back to Line 1.

Patience begins with 'Pacience is a poynt, þaȝ hit displese ofte'

and ends with 'þat pacience is a nobel poynt, þaʒ hit displese ofte'—
an even more pointed circularity.[99]

It is a pity that neither *Pearl* nor *Patience* shows the kind of overt
evidence of 22 which *Sir Gawain and the Green Knight* does. The latter
is more subtly circular in format. It does not begin and end with the
same lines but instead returns to the same idea: referring to a British
heritage traceable first to Felix Brutus then ultimately back to Troy.

Barrett points out the 'traditional ring structure' in those stanzas
describing the pentangle, a circularity that pointedly repeats the
name of the device on Gawain's shield and its royal red and gold
colouring. This, of course, can be expected since the poet was also
intent on emphasising the pentangle as an 'endless knot'.[100]

Another kind of circularity is that shown in the 12 illustrations
accompanying the manuscript. Maidie Hilmo argues that, although
they have been damned as crude and rough, the illuminations show
a sensibility to contemporary religious debate about iconography.
Many manuscripts of the time have been defaced or mutilated,
their lavishly decorated illuminations cut from the folio pages. The
illustrations of the *Pearl* manuscript have, however, survived intact.
Devoid of gold leaf, peacock colouring or intricate borders, they
carefully avoid several contentious issues raised by Wycliff and
the Lollards while being sensitive in the way they deal with others.
Neither the Godhead nor the saints are illustrated in the traditional
way: the closest any picture comes to a representation of the Trinity
is the three pearls worn by the *Pearl* maiden, a departure from the
single pearl of the text.[101]

The illustrations link one to another through various repeating
motifs: flowing water or sea beasts or a king's court. Hilmo points out
the figure of the 'dreamer' who seems, like some medieval version of
Alfred Hitchcock, to be making cameo appearances from time to time.
By repeating part of a picture, the illustrations have a kinship with
Pearl which repeats all or part of the last line of one stanza in the first
line of the next.

Still while the 12 illustrations may be circular in concept and may
be more deeply tied to the poems than we realise, they do not give us

any clues about the meaning behind the metaphor of 22.

Despite the failure of *Pearl* and *Patience* as well to show mathematical signs of circularity as well as verbal ones, we will pursue this theme through *Sir Gawain and the Green Knight* in the hope it may shed some light on the rest. If a poem is deemed to be 'circular,' then its length can be considered as equal to the circumference of a circle. The circumference will be $^{22}/_{7}$ times the diameter or, alternatively, the diameter will be $^{7}/_{22}$ times the circumference.

Yet what can 'diameter' possibly mean in the terms of reference of a poem? To find out, it is necessary to start looking at some poems and see what happens at a 'diameter'. In order to do this effectively, it's a good idea to use a mathematical technique which is common in determining algebraic rules. It's simply a matter of looking for a pattern. To do this pattern–spotting, it's best to choose a poem—neither too big nor too small. My 'just right' poem is 133 lines, which is one line over a multiple of 22. Because in mathematics it's the ones that are slightly off what you expect in a pattern that reveal the most, this seemed to me to be a Goldilocks–choice.[102]

Summer Sunday (or *Fortune*) is an alliterative poem that has much in common with the poems of the *Pearl* manuscript. It is the story of a man who rises at dawn to go hunting. Separated from his companions, he crosses a stream, passes into an 'otherworld' and comes upon the mysterious figure of Lady Fortune. On her Wheel of Fate many kings sit, one regally enthroned, one long deposed, one awaiting his turn to ascend on high.

Here the poet has depicted with vivid personification the traditional medieval belief in Fortune. Widely regarded as a lament for either the death of Edward II (1327) or Richard II (1400), *Summer Sunday* has a theme that is in fact too general to be absolutely sure either monarch is specifically intended. And while there is not much about its themes that would not have been very familiar to a medieval audience, its broad brushstrokes are markedly similar to the poems of the *Pearl* manuscript.

Like *Sir Gawain and the Green Knight*, it has:

- a rich description of a hunt

- a rhyming bob and wheel construction at the tail-end of its alliterative stanzas
- a length which is very close to a multiple of 22 in length
- a crafted golden section

In fact, its mathematical construction emphasises the golden ratio which is one of the hallmarks of the *Pearl* poet's work. The dialect of *Summer Sunday* is from the West Midlands and it is usually considered to have been written in the late fourteenth century, which places its composition not only around the same time as the *Pearl* manuscript, but in the same locality. The immediacy and vividness of the hunt scene, particularly in the images of the animals at dawn, evokes the same sort of vivid word picture as found in *Purity* where the harts and the hares raise their heads to heaven, hoping to be spared the full onslaught of the flood. *Summer Sunday* is a delight but it is also deeply thoughtful, a rare mixture of sensibilities found similarly in *Sir Gawain and the Green Knight*.

As far as I can discover, there has never been any serious suggestion that it was written by the *Pearl* poet. This, despite the fact that, besides being a multiple of 22 (like *Sir Gawain and the Green Knight* and *St. Erkenwald*), it is also a multiple of 12 (like *Pearl*) and in addition it opens with a sequence remarkably reminiscent of that part of *Pearl* where the dreamer comes up to a river which forms a boundary with an 'otherworld'.

In *Pearl*, the jeweller cannot cross the frontier but waits to see what Fortune will bring. In *Summer Sunday*, however, the hunter fords the stream and spies Lady Fortune with her wheel. At exactly $^7/_{22}$ of the way into the poem, right at the metaphorical 'diameter', Lady Fortune places her hand to the wheel and starts to spin it.

Summer Sunday therefore uses 'diameter' to highlight the concept of changing fate. This, in fact, suggests that it would be a good idea (at least from my vantage point as a mathematically-minded educator) to dispense with the term 'diameter' and substitute for it instead 'double radian': technically, this emphasises the turning angle rather than the linear diameter.[103]

We're going to discuss it so often, let's dispense with the imposing mouthful of 'double radian' and call it a 'dourad' instead.

Does this concept of a dourad fit other poems?

In *St. Erkenwald*, we find the dourad position (Line 112) marks the return of Erkenwald from an abbey in Essex to London, only a few lines after his character has properly been introduced.[104] In *Sir Gawain and the Green Knight*, we find the dourad (Line 805) is where Gawain has just arrived at Hautdesert and is about to approach the porter.

While neither of these are the climax of the story as we would understand climax in modern terms, they are both turning points in the plot. They are changepoints, moments of transition between one part of the story and the next. In *Sir Gawain and the Green Knight*, Gawain has just crossed the stream into the 'otherworld' (exactly as the hunter in *Summer Sunday* did) and is about to encounter the agent who represents a change of fate.

Against this, a remark of JA Burrow may shed more light on the design. In his analysis of Fitt II of *Sir Gawain and the Green Knight*,[105] he suggests that the turning point of fortune is highlighted and indicated by an initial capital beginning Line 763. As this is divisible by 7, it is worth examining in more detail. It is just after Gawain sends a prayer to heaven and just before he spots the castle of Hautdesert and would indicate a wheel of size 2398 lines. On the very next line—Line 2399—'chaunce' or *fortune* is mentioned. It may therefore be that the main wheel within *Sir Gawain and the Green Knight* does not run its course from the poem's first to last line. Instead, this key word is used to delineate the full revolution.[106]

Now *Pearl* and *Patience*, as we have already noted, are not multiples of 22 lines in length. Nevertheless, both have a verbal circular format, so we should not allow their lack of 22 to stop us calculating the position of a dourad for each.

This, of course, does not show a neat fall onto a single line that a multiple of 22 does. For *Patience*, it is on Lines 168/169 and for *Pearl* on Lines 385/386. In *Patience*, these lines describe the cries of the sailors on the ship taking Jonah to Tarshish as they call on various heathen gods. A significant narrative point, certainly, but not really a changepoint. So this idea about the significance of 22 would be ambiguous at best, but for what happens in *Pearl*.

Hugh Holman argues in the following paraphrase (finishing on Line 386, the dourad position) for a translation which incorporates an image he believes is lurking in the background of the text—that of the Divine Potter.

Although with goodness you, unblemished and divine, spoke,
I am only clay and a botcher's mistake; for I was marred in the
potter's hands; yet I have happiness in the hope that through the mercy
of Christ and Mary and John, I may be reshaped upon the divine wheel
into something pleasing to His eyes.[107]

The Wheel of Fate may not be mentioned but, if Holman is right, a very similar concept is: the wheel of the potter.

So why isn't there an obvious wheel in *Patience*?

Moreover, since there is a hidden wheel in *Pearl*, why doesn't that poem keep to the convention of 22? Assuming, of course, there is such a convention.

And, if there is, why does *Pearl* mention *wyrd*—the old Anglo-Saxon concept of fate—at Lines 249 and 273, rather than Line 386? Why do both *wyrd* and *destyné*—fate and destiny—appear together on Line 1752 of *Sir Gawain and the Green Knight*, so close to the end of the manuscript that there's no chance whatsoever of finding meaningful mathematics in any change of fate at that point?

And, even more importantly, why does Lady Fortune appear on Line 129 of *Pearl*, completely off her assigned 'station' if this interpretation of the dourad is correct? It's one thing to suggest that the *Pearl* poet deliberately added a mere four lines to the entire manuscript to make it a multiple of 17 but it's another to ignore hundreds of lines.

In fact, I have no intention of ignoring any of these. I will come back to all of these apparent 'discrepancies' in *Pearl* at the end of the next chapter.

However, I do want to tackle the question of the positioning of *wyrd* and *destyné* on Line 1752 in *Sir Gawain and the Green Knight*. At first sight, nothing makes sense, even working any possible wheel backwards. If Line 1752 is a dourad from the end, then the wheel itself extends backwards to Line 54 of *Sir Gawain and the Green Knight*. This entails a general description of the youthfulness of Arthur's court, nothing more. There is no changepoint, no suggestion of the march of destiny, no hint of fate spinning or stopping its wheel.

It almost blows the theory out of the water entirely. However, a possible procedure for mathematical analysis suggests itself on the realisation that, since there are two words for fate on this line, a wheel within a wheel is a possibility. It's necessary therefore to find another word for *fortune* somewhere else.

Indeed a euphemism for it might be 'chance' on Line 2068. Here Gawain leaves the castle of Hautdesert early in the morning and bids it 'good fortune forever.'[108] From the next line to the end is 462 lines. Not only nicely divisible by 7, but suggesting a full wheel of exactly 1452 lines and bringing us back to Line 1079 of *Sir Gawain and the Green Knight*.

Now this really does have overtones of Gawain stepping onto a wheel of fate: just prior to this line he has been informed of the close proximity of the Green Chapel to Hautdesert and, sighing with relief and relaxing his guard, he makes the fateful decision to stay at the castle instead of pressing on into the wilds.

Now examining the possibility of a wheel within a wheel, let's assume the size of the smaller wheel runs from this very line—1079—to the line where Gawain leaves the castle and bids it 'good fortune forever'—Line 2068.

The size of the dourad here would be 315 lines which, working backwards from (but not including) Line 2068 brings us precisely to Line 1752 where both *wyrd* and *destyné* are mentioned. It's too neat to be coincidental, especially when this very line hints, as already mentioned, of a wheel within a wheel.

I believe that multiples of 22 were so common and so uniformly understood as signifying a Wheel of Fate that the *Pearl* poet felt it necessary to move beyond the medieval version of a cliché and give his work a fresh edge.

However, there are apparent anomalies. But before delving into them, I believe I need to introduce a different side of medieval verse even to those well-acquainted with it. Even those who realise the *Pearl* poet was engaged in a cultural and religious war do not see that he was fighting on two fronts simultaneously. His battle strategy was far from conventional, at least from the distance of six centuries.

Both his battlefronts involved the meaning of numbers: on the one hand, he was defending against neo-Pythagoreanism and on the other, he was attacking in the face of the inexorable advance of the Hindu-Arabic decimal system. It is apparent that, for him, both Pythagorean mysticism and the Hindu-Arabic system had a religious base opposed to Christianity.

The fact is the entire length of the manuscript is a multiple of 17, held in antipathy by ancient Pythagoreans. It also features a diatonic comma which, instead of pointing to Pythagoras, arises out of a mathematical token for forgiveness. Herein is a statement of unequivocal partisanship. It is a line in the sand, a banner proclaiming as Luther was to do in a later age: 'Here I stand.'

Could four writers really have got together to write a poem cycle to conform to this mathematical battle strategy? Once we see how extensive the operational front is, it will seem almost too much to believe that even one poet conceived it and had the creativity to make it work.

To return to our consideration of *Pearl* and *Patience*: they have a verbal circularity which is not backed up by the conventional arithmetic motif of 22—the token I have suggested signifies a mathematical wheel. This, I maintain, was sacrificed to a higher agenda.

The sacrifice, however, will turn out to be very superficial. 22 is obscured but it's not absent, at least as far as *Pearl* is concerned. Once we bring it out of obscurity, it will be apparent that Lady Fortune is precisely on a dourad 'station', suggesting that the apparent discrepancy in her position was a blatant clue to the mathematically-literate reader about the oddness of *Pearl*'s numerical design.

By not featuring 22 overtly, however, the poet did have an opportunity to infuse a new idea into the Fate motif. He altered its emphasis towards the wheel of the Divine Potter, the shaper of destiny, rather than that of the Potter's servant, Lady Fortune.

There is an ongoing shadow of this sense of God–directed destiny also in *Sir Gawain and the Green Knight*, where the changepoint comes directly after Gawain has sent up to heaven a desperate appeal for help. Almost immediately he spots a castle that appeared, its turrets and battlements as unreal as decorative paper cut–outs, its walls and towers mysteriously 'pitched on a prayer': the last word related to modern 'prairie' and a pun for both 'meadow' and 'prayer'.

Thus we can see that the same sort of motif which is present in *Summer Sunday* is present in *Pearl* and, to a lesser extent, *Sir Gawain and the Green Knight* and *St. Erkenwald*.

Geoffrey de Vinsauf in his thirteenth century *Poetria Nova* advocated constructing poetry in a circular fashion. Moreover, he suggested building it mentally first: '*the measuring line of his mind lays out the work, and he mentally outlines the successive steps in a definite order... Let not the poet's hand be swift to take up the pen, nor his tongue be impatient to speak; trust neither hand nor tongue to the guidance of <u>fortune</u>... Let the mind's interior compass first <u>circle</u> the whole extent of the material. Let a definite order chart in advance at what point the pen will take up its course, or where it will fix its Cadiz. As a prudent workman, construct the whole fabric within the mind's citadel...*'[109] This advice which alludes to Lady Fortune and explicitly mentions circular constructions became radically scriptural in the hands of the *Pearl* poet;[110] however it is not necessary for Geoffrey's maxim always to have taken the style of the dourad.[111]

It would be perilous to assume that any 22–multiple automatically means that a Wheel of Fate motif is in operation. It works much of the time, but neither the wheel nor a critical changepoint which seals the fate of the main character appears at the dourad for every single poem which is a multiple of 22 lines.

Conversely, it would be equally perilous to assume that poems which are not multiples of 22 or 11 do not have a dourad in play.

For example, *Sir Orfeo* just like *Summer Sunday* fails to be a multiple of 11 by just one line. However, at the dourad position, the most critical changepoint of the poem occurs: Heurodis is snatched away.[112] *The Wife of Bath's Tale* also fails by just one line to be a multiple of 11, but at the dourad position, the knight in her story turns homeward.

On the other hand, the dourad combined with the 22–multiple does work particularly well in several versions of *Sir Patrick Spens.* See, for instance, the change between the second and third line of the following:

> *The first line that Sir Patrick read,*
> *A loud laugh laughed he;*
> *The next line that Sir Patrick read,*
> *The tear blinded his ee.*

In another version, it occurs on the fifth line of the following as the fateful voyage begins:

> *Be it wind, be it wet, be it hail, be it sleet,*
> *Our ship must sail the foam;*
> *The king's daughter of Noroway,*
> *Tis we must fetch her home.*
>
> *They hoisted their sails on Monenday morn,*
> *With all the speed they may;*
> *And they have landed in Noroway*
> *Upon a Wodensday*

Indeed, it works so well that the suspicion sometimes aired that *Sir Patrick Spens* is an eighteenth century fabrication is unlikely. Yet for all this, 22 as a mathematical token of a changepoint governed by the Wheel of Fate does not *appear* to be significant in such poems as the alliterative *Morte Arthure* or *Divine Love.*

However, it should be noted that the dourad of *Pearl* and *Patience* combined ($^7/_{22} \times 1743$) is almost 555 which, counted backwards from the end of *Patience*, falls on Line 1186. This refers to a 'garlande gay'—perhaps not unlike the circular crown of Anne of Bohemia.

Nonetheless, while this arithmetic metaphor for the Wheel of Fate works well a lot of the time, ultimately it is unsatisfactory because it doesn't work *all* the time.

David Howlett has suggested a very likely alternative solution for the meaning of 22. He remarked[113] that it may derive from the Hebrew understanding that in the opening chapter of Genesis there are 22 created things (and thus 22 letters in the Hebrew alphabet as well.) Moreover the Old Testament was long considered to be composed of 22 books (the minor prophets all combined in one section together.) Thus 22 as an arithmetic metaphor might stand for a Hebrew conception of the 'whole of creation' or for 'completion'.

In addition, 22 points placed evenly around a circle have some fascinating mathematical properties: if each point is to be linked to every other point in order to create what is termed in mathematics a 'compass rose', it will be necessary to draw 231 lines. The aliquot factors of 231 (excluding itself) add up to the seventeenth triangular number, 153. To be honest, I would never have known there were any interesting features of 231, had I not wondered about its significance after noticing it occurs in the design of both *St. Erkenwald* and *Sir Orfeo*.

Richard McGough points out there is an ancient Jewish tradition that says God 'placed the Letters in a circle.'[114] It's a short step from there to linking each letter to every other letter and creating a compass rose.

It may or may not have happened in the ancient world, however its advent in poetry of the Middle Ages is certain. Indeed, beyond its geometric form, Geoffrey de Vinsauf in *Poetria Nova* discussed it in the art of creating poetry which he described as if it were a building under construction: 'a circular structure..., a common shape, often subclassified as a 'rose' or other sort of wheel, or...*mappa mundi*.'

Geoffrey here advocated charting the geography of the imagination. Like modern fantasy and science fiction writers, a medieval poet could work within the circles of the world to create his own small bounded universe. The creative process was one of discovery: 'To invent [make a poem] is to come into a knowledge of the unknown thing through the agency of one's own reason.'[115] Many medieval explorer–authors evidently saw fit to finish off the journey into their creations with a numeric reflection of the Creator's own statement of completeness.

22 is an efficient solution to a construction problem which seems to have vexed English poetry since the time of *Beowulf*. It simplifies the mathematics, while retaining a sense of an old metaphor for circularity. *Beowulf*, for instance, is 3183 lines. It has not even a hint of a multiple of 22 in that length. Yet, curiously the monster fights begin in Fitts that are multiples of 11 and, far more curiously, a circle of circumference 10000 lines has a 3183–line diameter. As Joan Helm points out, such a metaphor of circularity is appropriate in a poem so concerned with rings and ring–giving.

10000 is a number reeking of 'completeness',[116] but as far as poetic composition is concerned, it doesn't have the flexibility that 22 has as a metaphor. It's far too rigid to be an elegant token.

Another solution to the problem is that illustrated by *The Phoenix* as it appears in the *Exeter Book*. Helm points out that it is divided by eight large capitals into eight sections. The 'ratio of first capital and the third expresses a perfect diameter/circumference ratio'[117] of the standard approximation of $^{22}/_7$ for π.

Still another solution may be suggested by the construction of the Old Norse–Icelandic *Lilja*, conventionally dated to the mid–fourteenth century. This hundred–stanza poem in honour of the Virgin Mary has been generally ascribed to Eysteinn Asgrimsson.[118]

The first and last stanzas are repeated, thus showing a very similar concept to *Pearl* and *Patience* and, to a lesser extent, *Sir Gawain and the Green Knight*. Thomas Hill has shown moreover that *Lilja* emphasises circularity in terms of salvation history and the Biblical narrative.[119] In addition, however, he also demonstrated it shows a triangular positioning with regard to the events of the atonement of Christ.

Hill pointed out this is suggestive of an equilateral triangle inscribed within a circle, emblematic both of the Trinity and incarnation. Accordingly, the 'triangular circle' thus serves as a motif of the Incarnation of Christ, the joining of God and man and the irruption of heaven into the affairs of earth, through the agency of the Virgin Mary—to whom *Lilja* is addressed.[120]

Returning our thoughts to the manuscript Cotton Nero *A.x* and the cross and pentangle within a circle identified within *Sir Gawain and the Green Knight* by Condren, it is worth considering whether there is also a triangle hidden in it.

I can find none.

However, I can find evidence for a circle and a triangle within both *Sir Gawain and the Green Knight* and *Patience* combined. This triangular circle partakes of much more complex mathematics than *Lilja*.

One–third of the way through the two poems occurs at 1020.7 lines into the second half of the manuscript which falls right at the end of the Fitt 1 of *Sir Gawain and the Green Knight*, exactly on Line 490.

490 thus assumes even greater prominence: it's a meeting point of several ornaments of mathematics within the text.[121]

Lest we be tempted to think of this as mere coincidence, the author has emphasised the existence of the triangular circle in another way. The relationship between the length of *Sir Gawain and the Green Knight* and the total length of the second half of the manuscript reflects an aspect of a triangle to the corresponding aspect of a circle. The ratio of the perimeter of the triangle to the circumference of the enclosing circle is 0.82664884 which is the same as 2531 lines to 3062 lines. This indicates that the length of *Sir Gawain and the Green Knight* when compared to the length of second half of the manuscript exactly[122] matches the mathematics of a triangle inscribed within a circle.

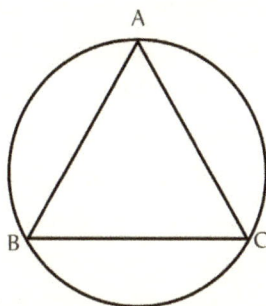

The circle represents the length of both
Patience & Sir Gawain & the Green Knight

The triangle represents the length
of *Sir Gawain & the Green Knight*

Circumference of circle = **3062** lines
Perimeter of triangle = **2531** lines
Radius of circle = 487 lines

Considering the circle:
A = first of Patience & last line of SGGK
B = Line **490** & C = Line 1510 SGGK

Considering the triangle:
A = first & last line of SGGK
B = Line 844 & C = Line 1688 SGGK

Thus, it would appear that the whole of the second half of the manuscript is governed by a mathematical metaphor for the Trinity and the Incarnation of Christ.

Yet that is not obviously the case in the poems themselves.

Of the many themes noted in *Sir Gawain and the Green Knight*, the Incarnation of Christ is not pre-eminent. In fact, as quite a number of commentators have noted, resurrection is far more obvious. Yet—and perhaps this might seem a pedantic point—incarnation is a necessary precursor to resurrection. You can't come back from the dead unless you are first born in order to die.

Still this leaves us with the unsolved puzzle. The length of *Patience* might necessarily be 531 lines simply to get the correct proportions for the circle vis à vis the triangle. Even so, why is *Patience* the most overtly circular poem verbally but not mathematically?

Of all the poems, it seems most obviously the one which should use a multiple of 22 in association with the Wheel of Fate[123] yet it is the one that seems most studiously to avoid it. After all, multiples of 22 simplify the metaphor of a circle considerably from the devices used in *Beowulf*, *The Phoenix* and *Lilja*. They would not only have been easy for a medieval reader to recognise (especially if each page is a multiple of 22 in itself as is demonstrably the case in some instances), they have the virtue of simplicity and unquestionable style. Unfussy and uncomplicated, as tokens of the Wheel of Fate or as motifs of Completeness, they would have provided endless choice to a poet.

They offered such a wide range of choice it is undeniably puzzling that *Patience*, as the most blatantly 'circular' poem of all, ignores them completely. Unlike *Sir Gawain and the Green Knight* where the last line can be judiciously put aside, *Patience* shows no sign of being amenable to 'completeness' or to the Wheel of Fate.

531 is admittedly just 3 lines away from a multiple of 22, but most significantly for this analysis, a changepoint does *not* occur at the dourad position of the poem.

Perhaps the solution to the problem lies in the fact that *Patience* is not complete by itself, that it needs to be looked at as just a cog in the wheel of the whole.

Or perhaps—much more importantly—it is also because, in this singular case, we've been looking at the wrong Wheel of Fate entirely.

digression 2

I'm not sure that I'll ever be able to feel completely comfortable with alphanumerics, mainly because of the connotation of gematria that goes with them. Like statistics, if you subject them to enough torture, they'll confess to almost anything you want.

Consequently, my heart as well as my mind leans towards mathematics, not numerology.

Nonetheless, because several commentators have suggested the *Pearl* poet uses numerology —admittedly, in some cases, simply reading 'numerical' as 'numerological', despite all cautions to the contrary—I have peered a little way into the gloom.

The good thing about this exercise was that it managed to uncover a few numerical features I'd previously overlooked.

Still I have come to the conclusion that there is no unambiguous support for the *Pearl* poet's use of gematria and plenty of evidence against it.

On the other hand, his use of numerical literary design is displayed everywhere. Mathematical metaphor is splashed from the first line of *Pearl* to the last stanza of *Sir Gawain and the Green Knight*.

Despite this, even 888, a well-known and common formulation in gematria for the name of Jesus Christ, fails to show up.

This is quite remarkable because the poet has ample opportunity to do so. It is doubly remarkable because I'm convinced he knew both Hebrew and Greek, both of which are alphanumeric languages and both of which naturally lend themselves to mathematical encoding.

It is triply remarkable because he had forgo some nifty

formulations to avoid common examples of gematria.

For instance, $153 \div (0.618)^7 = 4444$, which happens to be Line 889 of *Sir Gawain and the Green Knight*. There's a beautiful line–up of metaphors in that equation and the poet has used similar elsewhere, but nothing shows up in this instance.

I am convinced the poet was avoiding gematria.

Just like he was avoiding multiples of 10.

From the very beginning we noted he used 101 instead of 100. Sure, there is a very good reason to do so: it signifies the song of the angelic host and the Music of the Spheres. Sure too, there are very good reasons to have an extra 12 lines over the hundred in *Pearl* and *Purity* as well as an extra 31 over the hundred in *Patience* and *Sir Gawain and the Green Knight*.

However, the avoidance of multiples of 10 seems to be a deliberate and conscious decision on his part. This is remarkable because there is sweeping evidence for the poet's knowledge of the decimal system, both in its practical form and as a philosophic concept.

Nevertheless, watch out for the extreme rarity of multiples of 10 as we proceed.

Chapter 3

Angels Dancing: the number 0

As anyone acquainted with the history of mathematics is aware, prior to the arrival of the decimal system in sixteenth century Europe, zero as a concept had no opportunity to come into its own. Roman numerals were the order of the day and since they didn't have (or need) a placeholder, then zero simply did not exist in the world of medieval arithmetic.

Maybe someone should have told the *Pearl* poet he was in violation of the history of mathematics. I'm sure he'd have appreciated the irony.

A zero should, by its nature, be immensely difficult, if not completely impossible, to discern in the structure of a text. Fortunately, as I've pointed out previously, the poet was using a numerical literary artform, not numerology. So he didn't hide it; instead he took pains to point out exactly where it is.

Pacience is a poynt: patience is a virtue.

So begin most translations of the third poem in the *Pearl* manuscript, the 531 line story of Jonah.

Patience opens with a discussion of the Beatitudes. Like *Purity*, *Patience* is often regarded—at least by those who consider the *Pearl* manuscript to come from a single author—to be a work of the poet's youth. It seems to lack the maturity, the intensity of style and the sumptuous ornamentation of *Pearl* as well as the playful imagery of

Sir Gawain and the Green Knight.

Now, the idea that a poet does his best work late should have been discarded once and for all after the example of Wordsworth. However it is a remarkably enduring notion.

While I agree with those who consider that the four poems are all the work of one poet, I do not agree that *Patience* and *Purity* were the products of youth. The *Pearl* manuscript was clearly conceived *in toto*—no part of it can be excised without damage to the whole. This is demonstrated by the wealth of underlying mathematical metaphors which we are only beginning to uncover and which are so tightly woven together they only make sense as a single unit.

As mentioned at the end of the previous chapter, *Patience* is irritating: it is a verbally circular poem that lacks any mathematical motif of 'completeness'. It does, however, have a different arithmetic token: that of 'perfection'.

As Ed Condren noted, at 531 lines, it is a sum of so-called 'perfect' numbers.[124] If 1 is added to the first three perfect numbers—6, 28 and 496—the result is 531.

Now, would a medieval poet have picked perfect numbers just because they were 'perfect'? I really don't think so. Numbers weren't stripped winter-bare of their songs as they are today.

It's possible we should see 531 as a mathematical token signifying 'patience is the pre-eminent virtue and the sum of perfection'.

No doubt there is some sense of this in the structure.

However, I've found the poet to be eminently frank in telling us what he has up his sleeve when it comes to mathematics. So there are far more obvious metaphors I'd note.

For a start, since amongst the Beatitudes are Lady Mercy and Lady Purity, I'd like to notice numbers that are suggestive of the importance of their presence. Hence this 'sum of perfection' could also be considered as the result of adding 1 + 40 + 490.

The monad 1 is suggestive of an arithmetic metaphor for God. 40 suggests testing and purification.[125] Again we meet 490, indicative of forgiveness—its presence again perhaps telling us how significant the concept of mercy is as the lynchpin of the mathematical structure.

Now this set of associations—testing, purification, mercy—ought immediately to bring to mind the plot of *Sir Gawain and the Green Knight*.

In some ways, it should also remind us of the story of Jonah which forms the major part of *Patience*. Jonah is sorely tested, he is 'purified' in the sense he turns to doing God's will instead of running from it— and he famously complains about God's mercy. He rails against it, as if it compromises God's justice.

And with this, we can start to get a handle on why *Patience* features the story of Jonah rather than, as many commentators suggest would have been more appropriate, the story of Job who, after all, was famous for his patience. Once we look deeper into the structure we will see that *Patience* is not really primarily about patience. That just happens to be its first word.

It's about mercy, justice, faithfulness and peace.

Before looking more deeply at that, however, and its impact on the manuscript as a whole, let us return to the numerical symbols of *Patience*. I want to flag that *Patience* and *Pearl* together add to 1743 and then leave the discussion of this until much later. It is, however, an extremely important number connected to the zero about to be revealed.

The poet tells us the location of the zero twice, once at the beginning of the poem and once at the end.

Pacience is a poynt, he tells us at the start and at the finish: ... *pacience is a nobel poynt*.

The Middle English word 'poynt' is usually translated *virtue*.

Not everyone agrees with this interpretation, since the only other instance of it meaning *virtue* is in *Sir Gawain and the Green Knight*. The argument is so obviously self–referential that the possibility exists that in neither case does 'poynt' mean *virtue*. JJ Anderson suggests it is a word–play on *virtue* (or *necessity*) and *appointed*,[126] a far from unlikely possibility since puns abound in the four poems.

So what about a mathematical pun? Could *pacience is a poynt* also mean 'patience is a point'?

If we take it as such, it turns out to be an incomparable clue leading us through the labyrinth of the poet's dreamscape.

Geometrically, a point is of zero dimension. So the statement by the poet *pacience is a poynt* is an instruction to consider the entire length of the poem, all 531 lines, as being zero.

The remaining three poems in the manuscript are a total of 5555 lines—not merely a circular number in the Boethian sense[127] but also a multiple of that heavenly choir number 101, not to mention a multiple of 5 (so famously eulogised in *Sir Gawain and the Green Knight*) as well as a multiple of 11. A pity that it's not 22, but let's calculate the dourad anyway.

$$5555 \times {}^{7}/_{22} = 1767.5 \text{ lines}$$

What a delectable piece of mathematics! You've got to admire what happens next. Since *Pearl* is 1212 lines, this is exactly 555.5 lines into *Purity*.

He's actually resurrected the same set of digits through applying a motif for circularity.

Now $555.5 = 101 \times 5.5$ which brings up both 5 and 101 again. Moreover the specific line in *Purity* is *withouten maskle oþer mote, as margerye-perle*.

Apart from the overt reference to *Pearl*, indicating a link between the two poems, it is a statement of perfection (one of the mathematical tokens of *Patience*) and it is a changepoint. It is the very middle line in the transition passage between the story of the Flood and that of Abraham. It is a major turning point in the poem.

Unless this convergence of 5s, 101s and 11s is a coincidence of an unthinkable order, *Patience* is clearly designed as a zero.

Actually there's more.

We have noted that the 5555 lines of the three poems without *Patience* are resurrected in the 555.5 lines which emerge when the mathematical motif for circular completeness is applied. I chose the verb 'resurrected' in the previous sentence for several good reasons.

The date of Easter, the feast of the Resurrection, wheeled around in a regular cycle in the fourteenth century. Because it was calculated on a simple lunar–solar cycle, it repeated every 532 years.[128] Or to put it another way, after 531 years, Easter Sunday would come around again on exactly the same date.

So 531 or 532 are numbers that would automatically suggest

'Easter' to a medieval mind. Because they also imply a circular return to the same pattern, their resurrection motif is deepened. The idea of circular return which has been a key aspect of the poem from the start in the almost identical wording of the first and last lines becomes more explicable. The choice of Jonah also makes more sense in that it also alludes to the resurrection of Jesus:

One day some teachers of religious law and Pharisees came to Jesus and said, 'Teacher, we want you to show us a miraculous sign to prove that you are from God.' But Jesus replied, 'Only an evil, faithless generation would ask for a miraculous sign; but the only sign I will give them is the sign of the prophet Jonah. For as Jonah was in the belly of the great fish for three days and three nights, so I, the Son of Man, will be in the heart of the earth for three days and three nights. The people of Nineveh will rise up against this generation on judgment day and condemn it, because they repented at the preaching of Jonah. And now someone greater than Jonah is here—and you refuse to repent.'

Matthew 12:38–41 NLT

This changes the nature of the criticism sometimes levelled at the poet over his choice of Jonah as a biblical exemplar of patience. Although some commentators see Job who was renowned for this virtue in the face of adversity as a better option, the mathematics tells us that one of the fundamental themes is resurrection.

It is too easy to fall into the trap of viewing the poem through the lens of the name given it by modern editors. *Patience* might be its first word and it might extol that virtue during the first sixty lines (amongst many other virtues, it should be noted). However, the question is rarely raised as to whether the first word refers to the patience of the prophet or the patience of God.

The fact remains that the poem is untitled.

So, if one of its pre–eminent themes is indeed resurrection, rather than patience, then that should be evident elsewhere. Indeed the resurrection theme continues beyond the end of *Patience* and into the last poem of the manuscript, *Sir Gawain and the Green Knight*.

When the decapitated Green Knight picks up his head and rides off, he demonstrates a type of resurrection power.

Furthermore if we can consider the march of the seasons as a similar cycle of death and rebirth/resurrection, then perhaps it is no coincidence Michaelmas is mentioned 531 lines into *Sir Gawain and the Green Knight*.

The Feast of St Michael and All Angels, it was also a quarter-day,[129] it was the time for a fresh start: for hiring servants, paying rents, making removals, settling debts.

Perhaps, however, this is a little subtle, even though the poem has just been through a roll-call of the seasons. It may be stretching the point even though Michaelmas is mentioned as a pledge of the coming winter and a time when Gawain's mind began to turn to the vow he had made at the beginning of the year.

Regardless, it would appear that, as far as *Patience* is concerned, its 531 lines should be considered—at times—as also zero.

Now, this isn't just a breathtaking concept, it's mind–blowing. It's a staggering poetic motif in any age but, in the fourteenth century, when the idea of zero wasn't supposed to exist in Europe, it's just dazzling.

As we move further into the heartland of the poet's vision, it will become increasingly clear that his numbers do not make sense unless he was using the decimal system. This is not a man for whom any number was as good as another; nothing was arbitrary in his choices, so it is impossible in my view that he was not using some form of the decimal system.

Even prior to this stage, there have been some serious indications he had crossed the critical threshold: 5555 rendered in Roman numerals just doesn't have the exquisite economic fiveness of a decimal formulation.

$$\overline{\text{V}}\text{DLV}$$

The history of medieval mathematics is a morass of confusion and conflicting information. The advent of the Hindu–Arabic decimal system in the west is certainly not as clear–cut nor as late as many books state.

As early as the tenth century, Gerbert d'Aurillac employed Spanish monks to teach him the decimal system. Gerbert's learning was widely regarded with suspicion and he was accused of sorcery, but fortunately for Gerbert, this was end of the Dark Ages. A very tolerant view of witchcraft was then prevalent: it was regarded as a weakness in faith, not as an offence punishable by death. Ironically, as Jeffrey Burton Russell has pointed out, the heyday of the witch–hunts was the time of the Enlightenment.[130] In any event, the accusation did Gerbert little harm: he eventually rose to the papacy in the year 999 as Sylvester II.

A Spanish manuscript from the year 976 is the earliest European example of the distinctively curved and flowing Arabic numerals: the manuscript does not, however, include zero. Nonetheless by the late twelfth century, notaries included Hindu–Arabic digits in their documents; Raniero di Perugia for instance not only coded his work with such numerals, he naturally included zero in numbers like sixty.[131]

In 1202 *Liber Abaci* was published and its championing of the decimal system had an enormous influence on the intellectual climate. It was written by a merchant's son, Leonardo of Pisa, who has achieved enduring fame under the name Fibonacci. In 1299, the city of Florence banned the use of Hindu–Arabic numerals. It was thought that the system facilitated dishonest dealings because it was easy to modify a zero to a nine or a six. It was also thought that the place value system allowed deceitful merchants to inflate values by adding a new number to the end of a row. This seems to suggest that the decimal system was not confined to a few intellectuals, as is normally suggested. It was out in the marketplace and at least common enough to be a problem in Florence.[132]

Moreover the fourteenth century Provençal Jew Immanuel Bonfils of Tarascon was an outstanding mathematician who not only used decimal fractions with a notation possibly of his own invention, but also was capable of exponential calculus.[133] In fact during the late Middle Ages many Europeans rejected the very notion of a zero, regarding it as the creation of Satan.[134] By the early fifteen century the Middle English word 'nought' already carried the idea of numerical zero. It was seen as a 'distinctly uncanny figure, sinister or possibly embodying a sacred mystery.'[135]

So clearly the idea was around: it may have been a controversial notion but it wasn't unknown. In the late sixteenth century with the advent of the printing press, it became much more widely known. From that point, it never really looked back but it is nonetheless incorrect to suggest that the decimal system had not made substantial inroads into medieval Europe.

Not only is there no reason the poet could not have used the decimal system, he didn't need to be particularly avante garde to feature it. It had been around for centuries.

The poet has not been lackadaisical in laying a trail of evidence through his mathematical labyrinth. 'Poynt' is far from the only clue he gives us that he has used zero in constructing his poems.

The structure of *Patience* is disputed: some scholars see it simply as 531 lines, others envisage it as composed of 132 regular four-line stanzas, with one three-line stanza. With 132, we are back to a multiple of both 22 and 12.

However, it seems that to achieve this, we have to ignore that singular three-line stanza. In fact, it is often seen as so problematic that JJ Anderson, for instance, proposes that it is a scribal error. The poet, he suggests, had a tough time translating the Vulgate and wrote several lines of draft which the scribe was meant to ignore. Thus Anderson, along with others, considers the irregular stanza[136] to be an accidental inclusion.

It's not, however.

No question on that. Because, without it, *Pearl* and *Patience* do not add up to 1743. As we shall later see, this is a pre-eminent clue that there is, if not zero, then a vanishment of some kind somewhere in *Pearl* or *Patience*—or both.

Let us turn first to this 'irregular' stanza to see what it says for itself. The phrasing is obscure and difficult, but it contains an important clue in the opening line: *Bitwene þe stele and þe stayre disserne noȝt eunen*

The second last word is *noȝt*: nought. Nothing. Zero, by a modern arithmetic name.

Stele and stayre are generally translated as the riser and the rung of a ladder; an image of ignorance that is quite baffling. Why did the poet choose to say there are those who can't tell the difference between the rung of a ladder and the upright?

At this point we should remind ourselves that he is a man of many puns.

Siegfried Wenzel has indicate that 'stele' may not be the riser of a ladder as it is so often translated. 'Stele' and 'stayre' may *both* refer to the rung, or *both* be as different as a step and a stair! One rung and the next, one step and the next. Only the vertical between them.

Now how much horizontal space does the vertical consume? None at all. Perhaps there's really nought to choose between the stele and the stayre and perhaps that's the real secret of stele which, in yet another twist, seems to be a pun on 'secret'.

The secret of the stair is almost certainly that this irregular stanza is another possible zero. But a zero within a zero makes no real sense: it can only have effect as a zero when *Patience* is counted as 531. So, once more considering the totality of the manuscript, three lines subtracted from 6086 leaves 6083: we only need to remove *Hony Soyt Qui mal Pence* to achieve a perfect diatonic comma formulation.

Perhaps it's pedantic to stretch the point this far and perhaps doesn't really matter if the *noȝt* of this stanza is a clue indicating that, in general terms, there is *a* zero to be considered or whether it points to the fact that this three–line stanza is *the* specific zero in question.

As far as I am concerned, *Patience* itself is so clearly a null–construction that it is truly breathtaking. Even the conception of this poetic device, let alone its execution, is so daring and unusual it bespeaks a surpassingly creative mathematical mind.

Zero might have been seen as a work of the devil by his

contemporaries, but the poet wasn't content with merely defining the enemy. He set out to fight it.

He was right in thinking that zero is a religious concept.

It originates in the notion of nirvana. It is thus connected to samsara: the wheel of fate and re–incarnation in Hindu and Buddhist thought which can only be escaped when all desire is extinguished.

The circularity of *Patience* is truly connected to a Wheel of Fate. It's just not a European one. It is connected more by opposition than by agreement; the 'nothingness' of nirvana beyond all striving is utterly different to the 'nothingness' of perfection in *Patience*. Nirvana emerges from a life–denying philosophy that the world is illusory.[137] *Patience*, on the other hand, is wholeheartedly life–affirming.

The poet could have left the size of *Patience* as 531 and not also made it a zero. As 531 it is emblematic of resurrection but, as 531 and simultaneously zero, it is emblematic of resurrection in opposition to re–incarnation.[138]

What made him think of this idea? Does it come from the geometrical idea of a point shifted into an arithmetic framework? Or is it totally arithmetic in conception?

How did he look on his own handiwork and explain to himself, if not to others? How did he visualise a zero that is also 531? Did he think of it as the ultimate extreme of a recess construction? Did he imagine a host of 531 angels dancing and whirling on the point of a pin?

Today we might describe it as a galaxy trapped inside a black hole which has collapsed to a point or as a teeming world inside a microscopic water drop, but the poet didn't have that luxury. Did he think of *Patience* as a pleat in the fabric of the manuscript, a hidden fold[139] which could be smoothed out or re–creased at will? Or did he simply think of it as an allusion to the vision of Julian of Norwich—'I saw God in a poynte'—or to Dante's *Paradiso* where God is seen as a point of light around which seven concentric rings are revolving?

Even if he did, to transfer the concept to the length of a poem, is stunning.

To craft the lengths of two poems in such a way that the mathematics points on the one hand to resurrection (not re–incarnation) and on the other hand to incarnation (not re–incarnation) leaves me lost for superlatives.

The sociological impact of zero on the mindset of western civilisation is, paradoxically, not zero. Mental states are partly dependent on their ability to be verbalised or conceptualised.[140] In the Middle Ages, serfs and servants might have looked up to their betters and felt small by comparison, but they didn't feel like less than nothing. The introduction of the notion of zero–as–nothing along with negative numbers supplied names and ideas that could be applied to mental states. Zero is not merely a mathematical advance, it is also the first crack in what was to become a philosophic chasm.

To fully appreciate medieval literature, we have to cast off the mental chains of a post–modern culture that specialises in the literature of depression.

What appears as cheerful piety in comparison with the dark and gritty realism of modern literature and cinema isn't necessarily so—it is the natural gravitational centre of a culture in which the positive is a default state of being. This is not to say that gloom and doom never appear in medieval poetry, they do, but certainly not as nihilistic despair.

It's almost as if the *Pearl* poet took one look at the traditional nirvana–based zero and saw the future unfold. Hope replaced by desolation, negativity born in the mind of society, the seed of nothing crowned, absurdity feted. It would have been a nightmare vision to someone of his sensibility. In setting out to create a new metaphor for heavenly bliss which was a zero but not nothing, the poet set himself a task so daunting that the fact he actually achieved his goal is testimony to his genius.

With *Patience*, he was fighting a religious mind–war on two fronts and his strategy was to attack both at once. On one front is the Hindu–Buddhist concept of nirvana; on the other, Christianity's most ancient enemy, Gnosticism. The poet has his work cut out for him, but he may well have been the only man in the last few millennia equal to the task.

In case you haven't realised it, I am in awe of his talent. I have taught mathematics more than half my life but I've never encountered thought like this. There have been times when I've been lost in his labyrinth but, generally speaking, he has marked his passage very well. Oftentimes when his trail–marking blazes have vanished, it's because a modern editor covered them over and cleaned them up.

This brings us to *Sir Gawain and the Green Knight*. It also mentions 'poynt' in the famous passage on the meaning of Gawain's coat–of–arms.

Now, if the 'poynt' of *Patience* carries with it a connotation of a zero dimension, what about those 'poynts'? Are they exempt?

Surely the poet's unique conception of what a 'poynt' entails should apply in both places. The description of Gawain's shield where the 'poynts' are so important is 51 lines (3×17) in length and, by the poet's own admission, is a digression which will hold up the story:

I am in tent yow to tell, þof tary hyt me schulde (Line 624)

Perhaps this statement is an overt hint that these lines are to be regarded, like *Patience*, as zero. Perhaps the repeated use of 'poynt' within the digression is confirmation of its null–status. The poet may tease us with obvious clues or tantalise us with subtleties, but he never fails to mark the way.

One of his mathematical teases is the verbal—but not arithmetic—circularity of *Pearl*. At 1212 lines, it's not a multiple of 22. Nonetheless, as already noted, it echoes with the image of a Divine Potter at the dourad. Perhaps, however, an important clue has been overlooked. If *Patience* can hide a zero, then perhaps *Pearl* can hide a multiple of 22.

Of course once the poet had conceived of a 'zero' which could be either 531 or 51 or 3 lines in length, there was nothing to stop him changing its size again. Indeed Nick Davis points out the possible significance of Lines 273–276 in this regard:

> *And þou hatz called þy Wyrde a þef,*
> *Þat oȝt of noȝt hatz mad þe cler;*
> *Þou blamez þe bote of þy meschef;*
> *Þou art no kynde jueler.*

He suggests that here the poet conceives of God Himself

controlling the absolute value of what we would consider 'nought':

> *The pearl of inestimable worth that stands in the middle of the maiden's tunic might be considered to evoke the form of the numerical zero, the 'noȝt' already paradoxically endowed with a value ('oȝt') which God alone determines...*[141]

The *oȝt*–graced *noȝt* of Line 274 is connected to Wyrd—*fate*—in the previous line. Subtle as it is, it may well be a hint that there is a wheel of fortune operational at this very point.

Line 273 can be evenly divided by 7. This suggests it might be positioned at a dourad position of the full length of the poem. It does not, however, fit properly; it corresponds to a length of 858 lines which is, unfortunately, 354 short of the total. 354 is a number of considerable significance in *St. Erkenwald*, but why the number of days in the old synodic year suddenly pops up out of nowhere in *Pearl* will have to be shelved at this point as we return to what is truly obvious in an effort to solve this mystery.

The most distinctive technical feature of *Pearl* is its *concatenatio*—the linkage of one stanza to the next by the repetition of the last line of each stanza in the first line of the following.[142] The stanzas are strung like pearls along a necklace, knotted at each of the concatenation points. 101 stanzas would normally imply 100 examples of *concatenatio*, but because of the circularity in which the end echoes the beginning, there are 101 examples of a repeated line.[143] We're back with the heavenly choir again and rightly so, because if those 101 doubled lines are collapsed into 101 single lines, then the resulting poem is 1111 lines (11 × 101) long.

It may not be a multiple of 22 but it is suspicious enough to be a substitute. If this hidden multiple of 11 works, then it is also an affirmation of the 5555 lines already noted as the total when *Patience* is counted as zero.

Curiously, when we start to look at potential zeroes within the text, some discrepancies we've noted previously can be explained.

Lady Fortune, for instance, appears in *Pearl* on Line 129, nowhere near the dourad of the poem—or, for that matter, anything else.

However, once zeroes are considered as a possibility, then she points to Line 480, the very end of Section VIII as the overall size of her wheel. This suggests we might look for smaller wheels within the overall circularity.

An additional confirmation for the existence of smaller wheels (containing zeroes) is found if we return to a consideration of Lines 273 and 274. Line 273 mentions *wyrd,* 'fate' and Line 274 mentions *noȝt,* 'nought'. If we take this as a clue to apply one zero per stanza up to this point, Line 273 would effectively receive a new numbering of 251, which may explain the unusual positioning of another instance of *wyrd* on Line 249, just two lines earlier.[144]

There is little doubt that there are 101 zeroes carefully knotted into *Pearl*, making its length as much 1111 as 1212.[145]

1111 turns out to be hidden not only within *Pearl* but within *Purity* as well. Using 1111 as the size of an internal Wheel of Fate within *Purity*, the dourad would be 353 lines—precisely one line more than entire sequence featuring Noah and the flood. Also precisely one line more than *St. Erkenwald*—suggesting that 352 is a very significant number. Or one line less than 354, suggesting a 'triplet' of immense significance.[146]

354 is, in fact, exactly two-thirds of 531. Perhaps this suggests that *Patience* should be considered as having yet another geometrical formulation: that of an equilateral triangle inside a circle, just like the second half of the manuscript. This would also give it similar features to *Lilja* and suggests that at least three of the four poems are governed by an incarnational theology affecting the mathematical design of the whole manuscript.

The heavenly choir singing 'Peace on Earth' and heralding the birth of Jesus are indicated through the factor 101 featuring in the 1111 lines within that internal Wheel in *Purity*.

Clearly there are 101 infolding lines within *Pearl*.

This returns us to the question of how many zeroes are in the text. Are there other singularities which like black holes collapse the space around them? Is the well-known loss of a day in *Sir Gawain and the Green Knight* due to the poet's oversight or scribal error in leaving out a line? Or is it a hint about the existence of these zeroes?[147]

Should, for instance, every line which mentions 'poynt' be considered as zero? What about the two stanzas with the description of Gawain's shield which the poet calls a digression and to which the discussion of 'poynt' is central? Should the whole stanza or just the line where 'poynt' is featured be collapsed?

To take the idea of singularities a step further, should we consider the 'bob and wheel' at the end of each stanza in *Sir Gawain and the Green Knight* as yet another 'zero'? After all, Clifford Peterson noted that some are not in strict order: they are set to the right of the text and overlap the line numbering at various stages. In fact, he developed his theory that the author was J. Massey in part because Line 101 contains the word 'mas' and has the beginning of a 'bob and wheel' set alongside it, not after it.

For many critics, his argument seems a little tenuous. Rightly so. Perhaps the real reason the first line of the 'bob and wheel' abuts the text of some stanzas, rather than follows them, is because this is a hint to take these lines as 'zeroes' too. This sort of reasoning may seem a little tenuous as well, but there is no harm checking to see what happens if every clue, no matter how vague, is taken into account.

First, the 'bob and wheels'. There are 101 of them and each of them is 5 lines long. Subtracting 505 from the total of 5555 lines leaves us with 5050 lines.

The fact that this is a multiple of 101 is, at this stage, neither here nor there, because subtracting 5×101 from 55×101 will automatically produce a promising result with respect to 101.

However, 5050 is intriguing in its own right. It is the 100th triangular number. This means that it is the sum of every whole number from 1 to 100 inclusive. The 10th triangular number is 55 which we've just seen as a factor of 5555.

Triangular numbers are 1, 3, 6, 10, 15, 21, 28, 36, 45, 55 and so on. Their name comes from the fact they can be drawn as a triangle of dots and they have been long regarded as sacred—the Pythagoreans, for instance, called 10 the 'tetract' or 'tetraktys' when it was configured as a triangular number. The tetract was seen as Manifest Deity, Truth Incarnate and the Number of Numbers.

It is surprising that we haven't come across more triangular numbers in the architecture of the *Pearl* manuscript: not because of their Pythagorean background, but because of their Christian one. The opening hymn to the Logos in the Greek text of John's gospel is structured using triangular numbers[148] and Paul of Tarsus used them repeatedly as a literary construction device, particularly in his letter to the Corinthians.[149] We need to keep an open mind about whether the poet knew this, just as we need to keep an open mind about his knowledge of the decimal system. While we're here, it should also be noted that we need to keep an open mind about his knowledge of Hebrew as well.

If *Pearl* is considered as 1111 instead of 1212 lines, we should correct our previous figure of 5050 to 4949, a number which is reminiscent of 490, the token of forgiveness we noted earlier. In fact, if we now subtract the remainder of the digression about Gawain's shield, there are 4913 lines. If then all the lines where a singular noun for 'poynt' occurs, we're left with precisely 4900 lines.

It's a bit of a fudge, since we have to ignore a verb and a plural, but it is still far too neat for coincidence.

If almost all the possible 'zeroes' are accounted for, the length of the manuscript is exactly 490 × 10: the token of forgiveness times the number of commandments.

I'm inclined to think, however, that those last few references to 'poynt' should be counted and not dismissed. Most of them seem unlikely to be puns and, in addition, I am uncomfortable with having to make a special plea to accommodate the verb and plural form.

In fact I struggled so much trying to decide which of these last 15 'poynts' should be 'in' and which 'out' that I decided to check on the factors of 4913. And that made the situation even more complex because 4913 is 17 × 17 × 17.

It's difficult to decide at this point how many zeroes are in the text and to divine the poet's intention since 490 and 17 are so closely linked throughout the manuscript, as particularly evidenced in *Sir Gawain and the Green Knight*. Its first Fitt is 490 lines in length. The next line, Line 491, the beginning of the second Fitt, is 238 × 17 lines from the beginning of the text.[150]

17 is integral to the poet's 'mathematical signature'. It's a number for which he clearly has a special affinity.

However, regardless of which was in the poet's mind in this particular instance, either 4900 or 4913 affirms the existence of zero throughout the manuscript. In fact, the set-up is so meticulous, it is worth trying to discover the effect of those last two references to 'poynt'—the plural and the verb form—if we count them in as zeroes as well. They reduce the line numbering to 4898 which may not be particularly neat but is equal to 62 × 79.

62 is not only found as the extra lines—Bede's 'left hand' numbers—over the even hundred in the second half of the manuscript (31 'superfluous' lines in each of *Patience* and *Sir Gawain and the Green Knight*) but it is also the closest whole number to a golden section of 100. The golden ratio is exceedingly important to the poet's overall design so its presence here is quite in keeping with the mathematical motifs already under consideration.

Whether or not 79 has any independent significance is difficult to tell, however it may be a climacteric number. Climacteric numbers were generally multiples of 7 or 9 (although in this case, as in Shakespeare's sonnets, 79 is a combination, rather than a multiple) and were considered, even as far back as the time of Plato, to be those ages of a man's life which held particular significance and in which events of dramatic import were more likely. The possible existence of such numbers in the text points to the poet's ongoing concern with Platonic philosophy.

4900 is, however, in much closer focus than any of its neighbours and it brings us back to the 'defining number', 490. Moreover, it presents a new and interesting challenge with respect to the general perception of *Pearl*. There is a tendency to think of the first poem of the manuscript in terms of one special number: the dozen, the 'long' ten, the number 12. *Pearl* has stanzas of 12 lines and multiple references to 12: gates, jewels, layers.

Yet 1111 is clearly as important to *Pearl* as 1212 is.

Those intermediate results 5555 and 555.5 are obviously significant, not in that either is a multiple of 101, but in that one is 5 times 1111 and the other is half of 1111.

However, if 101 is a mathematical metaphor for the music of heaven, 144 for the community of heaven and 490 for the forgiveness which descends from heaven, what aspect of heaven does the metaphor of 1111 stand for?

It's time to interrogate Gawain and a quartet of ladies. Gawain in particular has a few questions to answer, as much about what he doesn't do as what he does.

digression 3

The tetraktys

Forget the pentagram. It was the tetraktys was the ultimate mystical symbol of the number–adoring Pythagoreans: 10 dots arranged in a triangle.

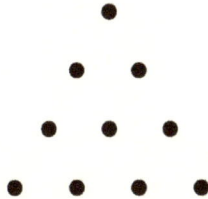

```
        ●

      ●   ●

    ●   ●   ●

  ●   ●   ●   ●
```

The tetraktys was regarded as Manifest Deity, the source of nature, the Number of Numbers, the Meaning of Meaning, the creative principle, the fundamental Truth of the universe, the heart of the Logos.

And by Logos, they didn't mean Jesus of Nazareth identified as the Word of God, they meant the divine Pythagoras who was an avatar of Python Apollo of Delphi and who had been re–incarnated numerous times throughout Greek history.

As I mentioned previously, the *Pearl* poet doesn't use multiples of 10 all that often.

490 is an exception—in fact, it may be the one and only case of it in the entire manuscript, other than possibly 4900. It's almost proving the rule because, of course, there is a very good theological reason for its use.

It seems that the poet is avoiding use of 10 and favouring use of 17.

10 is the number of the tetraktys, the focus of the adoration of the Pythagoreans.

It is also the base system of the Hindu–Arabic system with its peculiar and special philosophy of zero. It might feel normal to us now, having grown up in a culture in which it is familiar, but the poet's idea of nought is a fecund one, full of life, potential and possibility. It feels like a precursor to calculus. In complete contrast, ordinary zero is not simply empty—that would imply a vessel which had been vacated—and a vacancy is *not* nothing. It was an emptiness from which even the possible existence of a vessel had been removed.

These two theological reasons are why, I contend, the *Pearl* poet avoided the use of multiples of 10 as much as possible.

Chapter 4

The City of God: 1111

Allow me to introduce you to a delectable number (a curiosity, by the way, which makes its maximum impression in the decimal system): 12345679.

Note the number does not have an 8.

If we multiply by the missing 8, however, a different number disappears: $12345679 \times 8 = 98765432$

It becomes even more intriguing when multiples of 9 are involved:

$12345679 \times 9 = 111111111$
$12345679 \times 18 = 222222222$
$12345679 \times 27 = 333333333$
$12345679 \times 36 = 444444444$
$12345679 \times 45 = 555555555$
$12345679 \times 54 = 666666666$
$12345679 \times 63 = 777777777$
$12345679 \times 72 = 888888888$
$12345679 \times 81 = 999999999$

Even more fascinating is what happens when we apply 'circularity' by linking the very first number in the sequence above to the very last:

$12345679 \times 999999999 = 111111111^2 = 12345678987654321$

Truly that last number is the alpha–to–omega of the number world, at least as far as the decimal system goes.

While the 1111 of *Pearl* isn't the same string of ones as 111111111

81

by any means, it's certainly the obvious practical truncation, given the manuscript's total length.

Testing this speculation—that 1111 is simply a miniature of 111111111—is not especially difficult. It does raise major problems, however, with the received history of mathematics.

The *Pearl* poet was by no means original in the use of this numerical theme: Chrétien de Troyes had employed it some three centuries earlier in *Erec et Enide*.

As Joan Helm has shown,[151] Chrétien used it as part of a fusion of numbers which symbolised the mystical marriage of the earth and the moon, the union of heaven and earth, perfection personified.

How original Chrétien was has to remain open.[152] Far more important, however, than 1111 to Chrétien's poetic set–up was 1742.

On this very line, as Helm points out, King Arthur refers to heaven touching earth.

Helm sees 1742 as a truncation of 17424.[153] If it is, then it is just as appropriate to round up as to round down in the truncation.

As it happens, *Pearl* and *Patience* together total 1743 lines.

Together with the hidden 1111 of *Pearl*, this emphatically suggests the poet was working on a theme similar to Chrétien's in *Erec et Enide*.

This Arthurian romance was written in the latter half of the twelfth century.

Now this date is not too problematic as regards the history of mathematics. Leonardo of Pisa produced *Liber Abaci* about four or five decades later, but Gerbert d'Aurillac was more than 150 years prior to that time, so we're skating on thin ice with a decimal system at this stage but it's not out of the question.

What might be mathematically problematic, however, is Chrétien's source. Joan Helm argues that he got his idea from the book of Ezekiel.

The remainder, 5000 cubits in width and 25000 in length, shall be for common use for the city, for dwellings and for open spaces; and the city shall be in its midst. These shall be its measurements: the north side 4500 cubits, the south side 4500 cubits, the east side

4500 cubits, and the west side 4500 cubits. The city shall have open spaces: on the north 250 cubits, on the south 250 cubits, on the east 250 cubits, and on the west 250 cubits. The remainder of the length alongside the holy allotment shall be 10000 cubits toward the east and 10000 toward the west; and it shall be alongside the holy allotment. And its produce shall be food for the workers of the city. The workers of the city, out of all the tribes of Israel, shall cultivate it. The whole allotment shall be 25000 by 25000 cubits; you shall set apart the holy allotment, a square, with the property of the city.

Ezekiel 48:15–20 NAS

Here is the description of the City of God, 'Yahweh Shammah'— *THE LORD IS THERE*—the New Jerusalem which in the book of Revelation is the place of the Lamb and the brides of Christ. The New Jerusalem is, of course, not only a major iconic theme of *Pearl*, it is illustrated in one of the accompanying miniatures.[154]

By examining verses 16 and 17 in particular, we can see where 1111 comes into its own. The city is a square of side 4500 cubits. With the outlying suburbs taken into account, 250 cubits in each direction, the city plus suburbs is a square of side 5000 cubits. Now the size of the cubit isn't relevant where a ratio is concerned.

$$5000: 4500 = 1.111111111111: 1$$

In fact, the ratio involves an unending repeating decimals but if we consider it to three decimal places (or four 'significant figures'), it is

$$1.111: 1$$

This suggests that the hidden 1111 of *Pearl* is a mathematical metaphor for the City of God.[155]

Indeed the circularity of the poem alone or the immaculate roundness of the pearl has suggested to a number of writers a motif of endless perfection appropriate to a symbol for the kingdom of heaven.[156]

Moreover, perhaps the apparent discrepancy of 354 lines between the size of the 'Wheel of Fate' and the total length of the poem can now be resolved. The Temple inside Ezekiel's City was a square of side 250 cubits. Its diagonal would therefore be 354 cubits.

Taking into account now that

- *Pearl* is 1212 lines (12 × 101)
- *Pearl* plus *Purity* is 3024 lines (144 × 21)
- *Pearl* plus *Patience* is 1743 lines

we can see the mathematical metaphors beginning to link up.

I have already suggested that 144 is a token for the community of saints and that 101 is a token for heavenly harmony and the Music of the Spheres. What could possibly be more appropriate to connect them than 1111, a numerical token for the City of God—heaven itself—which has 12 gates of pearl and 12 jewelled[157] pavements?

Inside that 1111 there is a subset of 354, a numerical token of the Temple inside the City.

It is no coincidence that the *Pearl* poet is using a similar numerical structure to Chrétien's *Erec et Enide*. It is unquestionably deliberate and it is not a compliment.

JJ Anderson remarks on the allusion in *Patience* to the French *The Romance of the Rose* by Jean de Meun, suggesting that the *Pearl* poet was willing "to plunder whatever he could to make his point, for the *clene Rose* is not clean at all but a knowing *exposé* of the worldly *mores* of courtly society. In the passage noted in *Cleanness*, one critic observes, 'Jean ironically relishes the polished deceptions of love service.' *Cleanness* entirely overrides the original context of the passage, and forces it into the narrator's own mould."[158]

While I agree with Anderson that *Cleanness* (*Purity*) is indeed intended to override the original passage as *Pearl* overrides *Erec et Enide*, I disagree entirely that the *Pearl* poet was willing to plunder whatever he could to make his point. His targets were highly selective.

He wasn't engaging in theft at all; to describe a premeditated attack simply as 'plunder' is to miss the point. Don't be deceived by the chivalrous tone and the courteous demeanour. He might come with a holly branch in his hand like the Green Knight, but also like that jolly giant, his good manners veil ruthless intent. The poet's attitude is 'take no prisoners, leave no stone upon another'. He may indeed ignore the cynicism of the *Romance of the Rose* but that's to use it against itself. And is that not the supreme irony?

He was not the first to take on Chrétien de Troyes, Jean de Meun and the troubadours of France[159] but he had the potential to be the most formidable of all the enemies of courtly love. I do not think it too unrealistic to suggest that only the fact his work never seems to have been circulated stopped that potential from ever being realised.

Zealous in his loathing for the association of courtly love with magic and Pythagorean mysticism, he was able—with times of rare exception—to maintain a cool, assured detachment. Those times of rare exception, however, give him away entirely.

Joan Helm's dissection of the mathematics and Hebrew cosmology of *Erec et Enide* provides all the information required to explore the *Pearl* poet's response to Chrétien's neo–Platonism. She identifies 396, 1980 and 3960 as numbers particularly associated with Enide, 81 and 1056 as pertaining to Erec, while, as previously mentioned, 1742 as signifying the 'kiss of heaven and earth'.

Within *Patience* itself, there is a reference to such a kiss, though utterly different in its interpretation.

However before looking in more detail at those numbers Helm identifies and the 'kiss' planted by the *Pearl* poet, it is worth digressing again in order to examine briefly a notable absence from his work.

This absence may or may not have been obvious to any fourteenth century reader on the way through, however, by the last line it should have become suddenly and completely apparent. The chivalric wearing of a green sash[160] by Arthur's court to honour Gawain as well as the motto *Hony Soyt Qui mal Pence* should have abruptly alerted the reader to what *wasn't* there.

There is not a single tournament anywhere in *Sir Gawain and the Green Knight*.[161]

There's not even a joust at a ford or at a crossroad. Gawain meets dragons, giants, wood–woses, wild men and the most woeful of weather but he never tilts with another knight, not even during the Christmas games at the court of Arthur.[162] This is astonishing, given that so many tournaments of the late Middle Ages were called 'round tables' or Artushof specifically in order to evoke appropriate Arthurian connotations.[163] Elaborate scenarios were created at many tourneys so that the participating knights could act out lavish theatrical fantasies inspired by Arthurian romance.

Gawain of *Sir Gawain and the Green Knight* seems to have been just about the only chevalier in the whole of literary Christendom who kept the long–standing clerical ban on tournaments.

Apart from the fact that this suggests the poet has a churchman's respect for ecclesiastical law, there may yet again be an avoidance of all things even remotely smacking of Chrétien's influence: the most important of all civic festivals which involved a tournament was that of the 'Roys de l'Espinette' (*King of the Thorn*) or 'L'Epervier d'Or' (*The Golden Sparrowhawk*) held each year in Lille.

At first, the King of the Thorn was the winner of the tourney, but at some time during the fourteenth century, he came to be appointed by a college of former 'kings' thus bringing the format of an order of chivalry to the event. 'The main prize of the festival was a golden thorn, which seems to have been a symbol of Christ's crown of thorns. On the other hand, the festivals were firmly rooted in the literary tradition of the romances, and were even called the 'festival of the lord of Joy', a direct reference to *Erec et Enide*.'[164]

Enide in Chrétien's romance is the only one able to take the silver sparrowhawk from its perch. As Helm points out, the main number associated with her, 3960, has been well–known until very recent times[165] as the number of miles in the radius of the Earth. Thus Chrétien's use of 396 (one–tenth of 3960) and 1980 (one–half of 3960) serve as confirmation of the significance of Enide's status as a spirit of Earth measure.

The silver sparrowhawk is symbolic of the moon and the bird's perch of the lunar orbit. Again, it has to be asked: exactly how deeply hidden was this particular matter from Chrétien's esoterica? The good citizens of Lille seem to have figured it out: their alternate name for the Golden Sparrowhawk was the King of the Thorn. Christ's crown of thorns? Hardly likely given those two alternatives. They evidently knew Erec was a thinly veiled reference to the Man in the Moon[166] with his bundle of thorns.[167]

If the citizens of Lille had worked out what Erec represented, there is little doubt they also knew what Enide stood for. She is quite obvious by comparison. Chrétien actually pulled out some mathematical calculation for Erec as a representative of lunar measure; calculation that was unnecessary for Enide in her guise as the spirit of earth measure. For Erec, the reciprocal of 81 must be calculated. This is 0.012345679 (note it lacks an 8 and note also that the unwritten assumption behind Helm's analysis is that the decimal system was used by Chrétien throughout his romances[168]) and is found in a consideration of the City of Yahweh Shammah, *The Lord is There*, as described in the last chapter of the book of Ezekiel.

Let us examine the passage again.

These shall be its measurements: the north side 4500 cubits, the south side 4500 cubits, the east side 4500 cubits, and the west side 4500 cubits. The city shall have open spaces: on the north 250 cubits, on the south 250 cubits, on the east 250 cubits, and on the west 250 cubits.

Ezekiel 48: 16 – 17 NAS

The City proper is a square of side 4500 cubits. From this square, the suburbs extend 250 cubits north, south, east and west, so that it is now a square of side 5000 cubits. There is some argument amongst scholars as to whether the standard cubit of 17.6 inches or the royal cubit of 20.4 inches is meant. However, the beauty of a ratio is that it's completely irrelevant which type of cubit it is.

The ratio of the area of the City with suburbs to the city with suburbs is $5000^2 : 4500^2$ or 1.2345679:1, the digits reflecting the reciprocal of 81. Confirming that this is not simply a coincidence is the number 1056, also associated with Erec, and which in Scriptural context corresponds to the length of one of the angelic 'reeds' used to measure the heavenly City. These were six cubits in length, a cubit in this case understood as being the standard (and not the royal) measure of 17.6 inches. Thus the measuring reed used by the angels was taken to be 105.6 inches long and Erec's status as a spirit of

Heavenly measure is confirmed by the number 1056 associated with his name.

To wrap up the symbolism, Chrétien used the number 1742 to point out the incomparable harmony of the marriage of two such beautiful figures ('ymages'), a perfection in the case of Enide that so stuns the character of King Arthur that he exclaims she can only come from the meeting place of heaven and earth.

Helm shows that this number comes from the 17424 inches in the perimeter of an acre (one chain wide by one furlong long.)[169] Curiously related both to earth geometry and to the heavenly City of Yahweh Shammah, an acre of such dimension has a length and width that can be sub–divided into numbers that testify to the source of their inspiration.

> 12 lines = 1 inch
> 12 inches = 1 foot (= 144 lines)
> 3 feet = 1 yard
> 22 yards = 1 chain
> 4 rods = 1 chain
> 10 chains = 1 furlong = 40 rods
> 1760 yards = 1 mile

12 apostles, 144 thousand virgins, 3 persons in 1 God, 22 letters in the Hebrew alphabet, 4 evangelists, 10 commandments, 40 days in Lent (reflecting the number of days Jesus spent in the wilderness and the number of years the Israelites were wandering there) and 1760 inches in 100 standard Hebrew cubits. The numbers of the customary English measuring system are all just a little too evocative to be simply random chance, particularly when it turns out that, using standard cubits, the size of the City of God is an extraordinarily exact 1000 acres. If that wasn't enough, the square on the terrestrial radius of the Earth (considered as a perfect sphere) is almost exactly 10000000000 acres.[170]

It is too far–fetched to suggest such designer numbers just happened to emerge from the English measuring system.

Of the numbers connected with earth measurement, the *Pearl* poet has used 3, 4, 12, 22, 144 and 176 in various ways throughout *Pearl, Purity, Patience, Sir Gawain and the Green Knight* and *St. Erkenwald*. Even without Chrétien's use of a measure to symbolise the harmony of heaven and earth, it's probable that the *Pearl* poet would have built in these numbers anyway.

However, Helm has shown how pervasive the idea of the heavenly City was in medieval measurement. It was never a trivial matter in the Middle Ages (or even beyond it) to attempt to 'square the circle'. There were various interpretations of how this could and should be achieved; perhaps the most famous is Leonardo da Vinci's *Vitruvian Man* which shows his own 'take' on the problem.

Vitruvian Man showcases the ideas of Marcus Vitruvius Pollio, a Roman architect of the first century B.C., who applied the proportions of the human body to architecture and building. However, it was hardly a novel idea even then. Plato had combined language and number to define the 'world soul'[171] but it was Polyclitus of Argos who had translated this abstract Platonic (and Pythagorean) system of numerical perfection into tangible shape: he had created sculptures of a man and a woman whose very forms expressed perfection. So famous was this work that the qualities of Polyclitus' sculpture came to be regarded as natural symbols of earthly and heavenly perfection.[172]

The square inscribed in the circle brings us to the medieval vision of celestial perfection: the City of God was a square, but the moon and planets were believed to orbit in perfect circles.[173] Thus squaring the heavenly circle involved finding some alignment between the City of Yahweh Shammah and the orbit of the moon.

As it (almost uncannily) happened, this could be done with perimeters but not with areas.

The perimeter of the heavenly City of 1000 acres was exactly 316800 inches, while 31680 was considered to be the number of miles in the orbit of the moon if it was brought down from its perch to just touch the surface of the earth.

This serendipitous congruence of digits assumed even more importance according to Helm because it so happened that 3168 was also the numerological value of the Greek letters for Lord Jesus Christ, *Kurios Iesous Cristos*.

Helm quotes other important numbers in her analysis and includes other ways 3168 was used to show how the sphere of the Moon is brought down to touch the Earth, thus rolling up part of the heavens and eliminating the 'infernal regions' in between where the demons of the air were said to dwell. There was more than one way to bring about a union of heaven and earth using 3168.

However, the numbers I have pointed out so far are more than sufficient for the purpose of examining the *Pearl* poet's response to *Erec et Enide* and its occult marriage of earth and moon. Recapitulating, Chrétien's main structural numbers are: 81, 176, 396, 1056, 1742 and 3168.

Of these, the only numbers apparent[174] in the *Pearl* poet's work are 81, 176, 1056 and 1742, which he appropriately modified to 1743.

He has highlighted his use of 81 in *Pearl* so clearly that many commentators think the copyist has made an error. The beginning of Stanza 81 has an initial decorated capital which seems so out of place that Andrew and Waldron, for instance, have suggested that 'the scribe mistook this for the beginning of a new section' as a result of the extra stanza in Section 15.

A rare contrary view is that of Condren who points out[175] that this is the very stanza where the Dreamer and the Pearl Maiden are most closely in accord. He considers that this is a brief time of mystical union, a prelude to the Dreamer's vision of the descent of the New Jerusalem in the next stanza and his walk to the very threshold of heaven.[176] Only in retrospect can we see that, when the mystical union is achieved, the maiden speaks no more and quietly, unobtrusively, fades from the poem.

If Condren is right in suggesting this stanza indicates mystical union of heaven and earth, then it becomes a perfect opposition to Chrétien's mystical marriage of earth and moon both thematically and numerically.

However, the poet has not restricted himself to 81 as a numerical token of mystical union. The metaphor 1111 also implies this. The reciprocal of 81 is 0.012345679 which was also important to Chrétien and this, as we have seen previously, when multiplied by 9 yields the *Pearl* poet's 0.111111111, a numerical token of heaven.[177]

Let us leave 176 until a later discussion of the design of *St. Erkenwald* and turn to 1056. Recall that this is an allusion to the size of the 'angelic reed' used to measure the City of God in Ezekiel. The most obvious place to look for it is therefore somewhere in the description of the heavenly City in *Pearl*. Indeed Line 1056 refers to a river flowing from the throne of the City which is brighter than both the sun and the moon: *Watz bryȝter þen boþe þe sunne and mone.*

As JJ Anderson points out, this is a touch not found in the book of Revelation where the heavenly City is described: in fact, the New Jerusalem had no need of sun or moon since the Lamb was her light.[178]

The introduction of a lengthy element running counter to a significant theological point indicates how important this line is in the poet's thinking. Not only is its mathematical placement important so is its content. The poet really tries to draw the reader's attention to his interpretation of the mystic kiss of heaven and earth. He looks openly at the sun and moon, rather than veiling his symbols with the type of occult alchemical references that Chrétien used.

His allusions seem to indicate that he has interpreted *Erec et Enide* not as a lunar–terrestrial marriage, as Joan Helm does, but as the wedding of the moon and sun. The numbers she has identified clearly transfer across into the *Pearl* poet's work but with a slightly different emphasis. Indeed in *Purity* on Line 1056, he makes it quite clear that he is changing the poetic meaning of symbols used by French romance writers such as the rose. In fact, on the very next line (1057) he actually mentions Jean de Meun under the name Clopignel. In a very careful piece of positioning, this line is 2 × 643 lines back from the beginning of *Sir Gawain and the Green Knight* (with Line 1700 being 643 lines back from its beginning) while forward 643 lines from the same location is the Five Wounds on the Cross, the main 'golden' motif of the Arming Sequence.

However it might be argued with respect to 1056 and 1743 that, since they are derived from linear measurement, there is no sure indication that the poet really does have Chrétien's celestial union in mind. Let us examine, however, where and how the *Pearl* poet uses the last number.

Notably he created this number by adding the line totals of *Pearl* and *Patience*: 1212 and 531.

Now although we have not considered the golden ratio to this point, it is one of the most significant structural elements yet to be examined.

The golden ratio of 1743 is 1077 and that precise line in *Pearl* refers to the planets, having just spoken of the circuit of the moon. The sun is mentioned two lines later. Thus, we have a definite reference to the heavens, if not earth, suggesting we are on the right track.

The golden ratio of *Patience* refers to the Four Daughters of God and thus obliquely to Psalm 85:10–11. *Mercy and Truth meet together, Peace and Justice kiss each other. Faithfulness springs forth from the earth and righteousness looks down from heaven.*

Here we have not only both heaven and earth but a kiss between them. We've arrived at the theological lynch–pin of the poet's inspiration.

As I hope to show in succeeding chapters, every poem in the *Pearl* cycle is designed to point to this specific line in *Patience* which refers to Mercy and Truth, Peace and Justice.

This is the centre of the labyrinth. If we follow the mathematically–minded pilgrim's 'path to Jerusalem', we find waiting for us at the heart of the maze the Four Daughters of God—Lady Mercy, Lady Truth, Lady Peace and Lady Justice.

Originally from a Hebrew concept, they were famous throughout the Middle Ages for the many performances which dramatised their debate on the paradoxical attributes of the Godhead.

With a touch of literary alchemy, the *Pearl* poet took a concept as common as mud and transmuted it to gold.

And it really was as common as mud. Hope Traver cites numerous variations on the basic theme, which arose around about the tenth century as a Jewish commentary on the creation of man.[179] In a Christian reworking[180] which changed the emphasis to man's redemption, it became massively popular across Europe. Not only was it extraordinarily widespread, it assumed many forms, including drama, poetry and song.[181] Vestiges of the concept lasted even into the twentieth century.

It would be far from accurate to suggest that the debate of the Four Daughters of God followed a formula; in fact, the 'daughters' weren't always female and they were not always Mercy, Truth, Peace and Righteousness. Nonetheless most of the classic presentations of the concept has both these elements. The virtuous sisters were at total loggerheads. One of them would threaten to storm out of Heaven, never to return, because God was failing to be God, if he didn't act justly (or mercifully or peacefully or righteously or truthfully or whatever the perceived imbalance may have been). She would be backed by a second sister who would stomp out in her wake. When concord is finally achieved (usually because God's Son calls them back to outline a plan of redemption that will simultaneously satisfy Justice and Mercy), all the sisters kiss and make up. Heaven and earth thus met in beautiful harmony.

The *Pearl* poet made use of this widespread concept as he took up arms against a worldview epitomised by courtly love. For him, the meeting place of heaven and earth was not to be found in Chrétien's esoteric measures with their disturbing neo–Platonic overtones and belief in re–incarnation. For him, the union of heaven and earth is in the moment of reconciliation between the Four Daughters of God: the kiss of peace and justice, the caress of mercy and truth.

As the Psalmist wrote, this is when heaven and earth meet, when the distance between them collapses to nothing.

Going back now to the 'zero dimension' of *Patience*, recall the questions I asked of how the poet thought about it in his own mind: *Did he imagine a host of 531 angels dancing and whirling on the point of pin? Did he think of it as a pleat in the fabric of the manuscript, a hidden fold which could be smoothed out or re–creased at will? Or did he simply think of it as an allusion to the vision of Julian of Norwich—'I saw God in a poynte'—or to Dante's Paradiso where God is seen as a point of light around which seven concentric rings are revolving?*

I think he saw it as a fold in the fabric, as a tuck in the scroll, as perhaps a reference to the opening of the epistle to the Hebrews where the heavens are rolled up: *Ipsi peribunt, tu autem permanebis, et omnes ut vestimentum veterascent: et velut amictum mutabis eos, et mutabuntur: tu autem idem ipse es, et anni tui non deficient.*

Let's say, for the sake of illustration, that 1 line equals 1 centimetre. Imagine the manuscript as a scroll or a straight piece of fabric 6086 centimetres long. Nearly half way along—3024 centimetres from the beginning—a fold is inserted. 531 centimetres are tucked up into a pleat. This means that, when the scroll is laid flat, the resulting length appears to 5555 centimetres.

Now if the fold is engineered to be symmetrical—perhaps it's actually envisaged as a kiss—the pleated section will be able to extend 265.5 centimetres in either of two directions. That is, it can touch as far as 265.5 centimetres forward into *Sir Gawain and the Green Knight* or as far as 265.5 centimetres back into *Purity*.

What do the lines which correspond to these positions actually say? Do they refer to the kiss of heaven and earth, to peace and justice, truth and righteousness?

Of course they do.

A symbol of peace features in *Sir Gawain and the Green Knight* while the coming of justice appears in *Purity*. The holly bob of the Green Knight—the evergreen symbol of peace—is mentioned in the former and the horror of the king of Babylon as he looks at a mysterious writing penned by a disembodied hand is detailed in the latter.

So the poynte of *Patience* is not only the collapse of the infernal regions between earth and heaven to zero, but a kiss.

The kiss of peace and justice, mercy and truth.

The great paradox of Christianity is that it proclaims a God of perfect love and a God of perfect righteousness. Mercy and justice are, humanly speaking, incompatible qualities—the only thing more important than justice is mercy and the only thing more important than mercy is justice. Yet to hold rigorously to one is to reject the other. A judge who is always clement denies justice to the victims of crime; a harsh judge who always invokes strict justice, never taking account of circumstances, is an equal failure.

The poet tries to keep a balance, as he moves from one extreme to the other. *Purity* is about God's righteousness and justice, while *Patience* is about his forbearance, mercy and pity: the two poems sit astride the middle of the manuscript, their combined total almost exactly the golden ratio of the whole cycle.

The viewpoints of the poems are so polarised that at times some approach diatribe. But this is perfectly understandable if the manuscript as a whole is a debate where the Four Daughters present individual rhetorical cases for their own virtues. Though perhaps it's better to suggest that this is not a dispute, since the poems only occasionally stray outside the discussion of their own particular virtue—*Patience* notably excepted. It's more of a presentation of a set of individual cases. JJ Anderson points out that the emphasis in *Purity* on righteousness is so extreme that Christ's ministry of healing is presented not as an exemplar of his compassion but as his distaste for uncleanness. Neither is there any reference to his death. However, this all makes sense if this is the case of Lady Justice—mercy is not in this equation, but is reserved for later discussion in *Patience*.

There are theological extremes in every poem[182] but *Purity* certainly is the most blatant. Since many people naturally favour justice and consider it to be the pre–eminent aspect of God's nature, it is likely the poet did this deliberately to show what a harsh judgemental world it would be if righteousness alone ruled God's character and his holiness reigned sole and supreme. Instead of seeing *Purity* as the poet's own thoughts on God's nature, I believe it should be viewed simply as an exercise in rhetoric.[183] Lady Justice is brought into the council of God and given an opportunity to make her best case.

The poet has taken her case to its logical limits and shown that justice by itself produces a God of terror and harshness. His technique is to start with the reasonable and move it by degrees to the extreme.[184] In *Patience* he shows that mercy by itself produces a presumptuousness on the part of mankind in response to a God of tolerance.

There have always been stories like this: stories where two extreme positions are juxtaposed. Readers tend to impose their own inclinations and cultural assumptions on such stories. One famous tale of extremes is a parable that Jesus told. It is generally called *The Prodigal Son*. The parable conforms to an ancient Jewish story–telling tradition: that of two sons, the eldest of whom receives justice and the youngest of whom receives mercy.

There's an almost universal tendency to overlook the first aspect of that customary set–up to the characterisation: justice is dispensed to the older brother in order to contrast his behaviour with that of the younger. Such a possibility reshapes the narrative intention considerably and, consequently, whether the story has been mis-interpreted or not is an open question.

I've always wondered why commentators go to extreme lengths to explain the significance of the minituae of detail in the story, including the fact that some words specifically suggest the whole village is invited to the celebration. Everyone is there to feast on the fatted calf. What's generally ignored here is that there is one exception to this invitation: the older son. No one has bothered to go out to the field to ask him in. He's still out working as usual and eventually he is revealed to be so out of the loop, he has to pull up a young boy and question him about what is the noise means.

The possibility that the older son is symbolic of a recipient of justice and designed in character to contrast to a recipient of mercy seriously impacts the usual interpretation of him as indignant, self–righteous and sanctimonious. If this parable of Jesus was meant to expose the qualities of mercy just as much as the qualities of justice, it's certainly no longer seen that way. It doesn't matter what age we live in—if we believe in God at all, we tend to lean towards only one perceived aspect of his character: justice, mercy, truth or peace.

A balanced view, which brings all of these virtues together, was the goal of medieval productions of 'The Four Daughters of God'.

In my view, it was also the goal of the *Pearl* poet. In *Purity*, as we have seen, he showcased righteousness and justice while, on the other hand, he featured mercy and lovingkindness in *Patience*.

Going further, in *Sir Gawain and the Green Knight* he demonstrates that faithfulness is a superb ideal but cannot hold true to itself when one covenant obligation is pitted against another. Gawain might be flawed in that he loves his own life enough to lie in order to save it—but there's also a church injunction against murder which includes self–murder. Avoiding a suicidal situation brings this complexity into play.

Lastly, in *Pearl*, the poet demonstrates that peace isn't a matter that can be easily settled even with a revelation direct from heaven.

And as we now move into an examination of the golden ratio within the text, it is fitting to note how well these themes fit both the four poems in the *Pearl* manuscript and also the virtues exemplified by Four Daughters of God.[185]

Almost every commentator has noted at some point that the theme of *Sir Gawain and the Green Knight* is truth or *trawthe*: faithfulness, fidelity, integrity.[186] Here again the rhetorical argument is taken to its extreme and exposed as wanting. By the end of the poem, it is made plain that truth must necessarily be allied with mercy. The human condition is such that, when obedience to truth also involves faith, honour and promise–keeping, inevitably there will come a time when one promise is pitted against another.

Gawain might be the 'parfait gentil knyght' but he also exemplifies the human condition for a man of integrity: honour is bound up with a man's faithfulness to his word but, sooner or later, life being as complex as it is, a conflict will arise between one oath and another. Gawain is trapped between the covenant he makes with the lord of Hautdesert and the promise he makes to the lord's wife. The covenant is not a legal contract to be broken if inconvenient; it is a solemn unbreakable vow.

Tolkien points out the Gawain was, in fact, trapped long before he undertook the covenant with Bertilak.[187] He was trapped from the moment he stepped forward, offering to take King Arthur's place and extricate him from a clearly dangerous blunder—participating in a 'game' that was hedged about with magic on the one hand and grave promises which echo covenant–like vows on the other.

Indeed Barraclough comments that the first covenant between

Gawain and Bertilak was given authority at Camelot by Arthur's invocation of a knight's investiture ceremony; the second was sealed by a drink.

Gawain's physical, psychological and spiritual entrapment deepens as the story progresses.[188] Facing certain death, he is offered hope: a 'magical' belt which may save his life. Although he considers it a 'jewel for the jeopardy', there is never a point at which he is completely confident of its efficacy. Significantly, the final trap laid for Gawain is not entirely directed at his carnal nature (which, after all, has withstood considerable assault to this point). It mixes in a spiritual factor, thus muddying Gawain's motivations: this factor is the command not to commit murder. It includes self–murder. Obedience to trawthe requires him not to place himself in a position of committing suicide; obedience to trawthe also requires him to eschew magic in favour of faith. The fateful moment when Gawain is offered the magic girdle has a long shadow extending all the way back to the moment he stands up in King Arthur's court, ready to save the king from his own folly.

By the end of the story, the poet has shown that, in a world where spiritual entrapment awaits even those of the utmost integrity, mercy needs to step in so that honour may be restored. Indeed, as Burrow points out, quoting *Ancrene Riwle*, the prospect of God's forgiveness should be held out but should be tempered, so that one 'neither despairs before God's justice nor presumes on his mercy.' Indeed, Burrow makes an explicit connection between the Green Knight's act of mercy and the Four Daughters of God, quoting Psalm 85:10, and commenting that the wide currency of the fable indicates a medieval preoccupation with a particular moral problem: how can anyone be both merciful *and* just?[189]

Yet the theme is not seriously underlined at the end of *Sir Gawain and the Green Knight*. At least not to the degree that trawthe has been given prominence throughout the poem. But then, it doesn't need to be strongly accented: it has already been treated to its own showcase in *Patience*, just as righteousness and justice are the featured highlights of *Purity*.

Now, clearly, if the Four Daughters of God do indeed have a poem each to present their case, this affects the interpretation of *Pearl.* Contrary to tradition, I contend that this luminous poem with its air of lofty serenity, [190] its elements of sleeping, dreaming and dying can only have had—at least in the poet's mind—one possible theme. That corresponding to the last of the Four Daughters of God: peace.

digression 4

Polygons inside Circles

The construction of *Patience* and *Sir Gawain and the Green Knight* is based around the mathematics of an equilateral triangle enclosed by a circle.

The presence of 1111 within the mix suggests there might be a square enclosed by a circle.

This all begs the question: is there any evidence for this?

Furthermore, since a regular pentagon, hexagon, heptagon, octagon and so on can also be enclosed by a circle, is there any evidence for them?

Any polygon, depending on its orientation, could touch a circle halfway along a circumference of 3062 lines. If there is any evidence, the halfway point would be a natural place to find it. As it happens, this is 1531 lines which, since *Patience* is 531 lines, is exactly 1000 lines into *Sir Gawain and the Green Knight*.

There Gawain is about to sit down to dinner with the Lady of Hautdesert and, unknown to him, his enchantress aunt, as well as his 'nemesis', the Green Knight. Nothing is obvious.

However four lines further on there is a key word, suggesting it would be advisable to take a closer look. That word means 'in the middle'. It's a pity it's not four lines back. If, however, it is a middle of some kind, then it would be necessary to excise the first eight lines of *Patience* to make it sit correctly. Actually, these first eight lines can be removed without the slightest difficulty or loss of sense.

But what affect does this have on the relationship between the new circumference (at 3054 lines) and the equilateral triangle found earlier? This triangle would now have a perimeter of 2525 lines—which, of course, brings up the multiples of 101 again. It is also, as Kent Hieatt pointed out, the number of lines from the start of *Sir Gawain and the Green Knight* to the end of its 'long' lines.

This seems to suggest that there are margins—five lines at the end, eight lines at the beginning[191]—and that therefore the poet designed this, just as Ed Condren suggested, using geometrical instruments.

Of course, Condren has already identified a pentagram and that automatically indicates the existence of a pentagon.

The pentagon has particularly interesting properties in that 3062 lines ÷5 = 612.4, making every vertex of the pentagon into a 'node' with 'lel letteres loken'. 612 is 4 × 153 while 613 is the number of commandments in the Torah, the law God gave to Moses. All of these nodes occur in *Sir Gawain and the Green Knight* which is particularly appropriate, given the thrust of Gawain's story is faith, covenant and truth.

There is also a possibility of a square and heptagon within the circle. The square construction based on 1.111 occurs at Line 2225 of *Sir Gawain and the Green Knight*. The slight difference between a perfect alignment with 2222 seems deliberate: this is 2 × 153 (or 18 × 17) lines from the end of the poem.

Line 1 of *Patience* ~ A ~ Line 2531 of SGGK

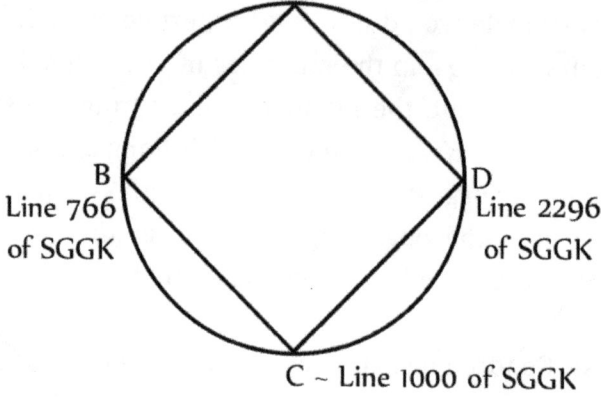

B
Line 766
of SGGK

D
Line 2296
of SGGK

C ~ Line 1000 of SGGK

For the heptagon, the ratio of polygon's perimeter to the circle's circumference falls on Line 2429 of *Sir Gawain and the Green Knight* or 102 lines (or 6 × 17) from its end. The hexagon, however, seems the most likely:

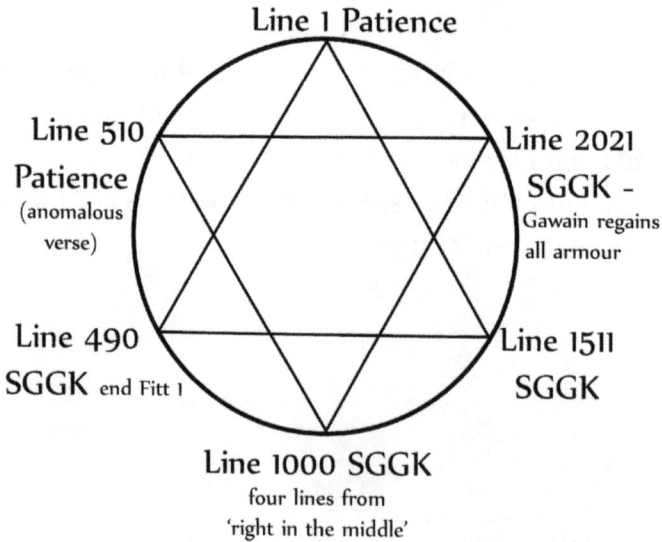

Line 1 Patience

Line 510
Patience
(anomalous
verse)

Line 2021
SGGK -
Gawain regains
all armour

Line 490
SGGK end Fitt 1

Line 1511
SGGK

Line 1000 SGGK
four lines from
'right in the middle'

Chapter 5

Gold and Peace: the ratio 0.618: 1

The library defends itself, immeasurable as the truth it houses,
deceitful as the falsehood it preserves. A spiritual labyrinth, it is also a
terrestrial labyrinth. You might enter and you might not emerge.

The Name of the Rose, Umberto Eco

Before tackling the golden ratio in the works of the *Pearl* poet, it's necessary to clear the ground around it. There are so many misconceptions and so much confusion about the golden ratio that even eminent mathematicians are apt to have foot–in–mouth disease when talking about it.

Let me begin by stating that the golden ratio is *not* phi. It is often confused with phi and the terms are frequently used interchangeably. However, phi and the golden ratio are actually reciprocals[192] of each other—at least 'big phi' (Φ) and the golden ratio are, while in fact 'little phi' (φ) and the golden ratio often (but not always) refer to the same thing.

Nevertheless, because unless otherwise specified, phi means Φ, this means that it is incorrect to suggest Φ and the golden ratio are one and the same thing. (Quite apart from the fact that a ratio, mathematically speaking, refers to the comparison of various parts of a whole, while Φ at approximately 1.618 is self–evidently more than a whole.)

This may sound like pedanticism gone mad, but unfortunately it's not simply a technical point; it has very important implications. If we impose our own limited understanding of phi on the past, then we will miss realising just how creative many poets actually were.

Phi is actually a growth factor of approximately 1.618 which, in the early years of the twentieth century, was introduced by William Schooling in a section of *The Curves of Life* by Theodore Andrea Cook. The idea was first advocated according to Schooling by the American mathematician Mark Barr. It is symbolised by Φ and draws its name from Pheidias, legendary architect of the Parthenon.

Now there are critics who assert that Φ has become some kind of mystical touchstone of architecture and that the claims for it are wildly exaggerated: they maintain that the Parthenon was not built using Φ, nor was the Great Pyramid, nor were selected stone circles in Britain, neither was Chartres Cathedral, the United Nations building or a multitude of other structures, both ancient and modern. Up to a point, this criticism is valid. Arguments about aesthetics tend to fixate on statistics or the eye of the beholder. The less said about either of these, the better. The problem is exacerbated by those people who seem to have the ability to spot the golden ratio almost anywhere.

Obsession seems to run two ways on this issue. Some critics demand inhuman perfection and impossible exactitude in measurement while others tolerate the fairly standard allowance for error in scientific circles: 2%. Outside of 2%, there is no doubt additional evidence is required. This is sometimes difficult to assess with any surety, as for instance, in the case of the Great Pyramid, where the statement of Herodotus is disputed.[193]

Personally, I feel a 2% allowance is quite excessive. I'd recommend keeping within 0.2% (that is, I would consider an error of 1 line in 500 to be the absolute uppermost limit of what is acceptable) and even that is often far too much.

If it is clear that a poet was using geometry as the basis of his design, then a single line error in a thousand is probably too much. On the other hand, if the design is clearly arithmetic in conception, then a leeway of 1 in 500 is quite reasonable.

Much as I dislike gematria and try to avoid it, if at all possible, it has one very positive aspect. The idea of a margin for mistakes is totally absent. That's not to say the answer has to be 'spot on' every time, 'touching' is good enough. Sometimes even two away from the designated answer is ok. However, beyond that is not permissible. The concept of scientific error simply didn't exist.

When I first started this investigation into Cotton Nero A.x, I was following the lead of Ed Condren in examining the golden ratio features. It was only as it became impossible to account for the discrepancies in the design—at one point just 1 line but often 2 lines, occasionally 4 or 8 and once a totally incomprehensible 20 lines out—that I began to consider the possibility of mathematical overlays. Sure the golden ratio was important to the poet, but not nearly as important as the numbers pertaining to the 'Kingdom of God' and the kiss of heaven and earth.

Thus, because of these slight discrepancies which have resulted from the complex mathematical weaving of different arithmetic metaphors, *Cotton Nero A.x* actually has a golden ratio design that is closer to those examples that occur in real life, than to any mathematical ideal. The Greek concept of perfect beauty using the golden ratio has been sidelined in favour of a more true–to–life approach.

Some additional important points on this matter need to be made before proceeding. There are many valid criticisms regarding the application of the golden ratio in practical situations. There are also many more unjustified ones: the reason some people are able to spot the golden ratio just about anywhere is because the tendency to *approach* the golden ratio really is everywhere in nature.[194] Theodore Cook in *The Curves of Life* listed a vast number of naturally–occurring examples of phi but, in reality, he was barely scratching the surface. The fact is, spotting a couple of daisies with 19 or 22 petals instead of the more common 21 and reporting that as a significant finding is an inordinate failure to understand what a *ratio* or what a *limiting tendency* is.

Another issue that arises regarding the golden ratio is the

provenance of the name. It is usually attributed to Martin Ohm in the nineteenth century who used the term 'goldene Schnitt'—*golden cut* or *golden section*—and the symbol τ. Ohm, however, made no claim to inventing it and his usage seems to indicate it was commonly known.

The name is often also said to have originated with the astronomer Johannes Kepler in the sixteenth century.[195] Yet Campanus of Novara was allegedly using the term 'sectio aurea'—*golden section*—in the thirteenth century and is said to have produced a very ingenious proof of its irrationality[196] which was much later emulated by the renowned mathematician, Georg Cantor.[197] Whether Campanus did, in fact, use the term has been questioned; however, the use of the word 'gold' to mark the golden section of poetry can be traced, as we shall see, from fourteenth century England to twelfth century France back into ninth century England.

Certainly the evidence is circumstantial, but I believe it is sufficient to be mindful of Henry David Thoreau's advice that 'some circumstantial evidence is very strong, as when you find a trout in the milk.' There's something extremely fishy going on with the terms 'golden ratio' and 'golden section' and while they may not arise in antiquity,[198] they are unquestionably older than generally admitted. I personally doubt their provenance is ninth century England, but I have not been able to find any evidence further back than that at present.

Last in our roundup of misconceptions regarding the golden ratio is a historical consideration. Medieval scholars simply did not know of phi—neither of 'big phi' Φ (approx 1.618) nor of its inverse, 'little phi' φ (approx 0.618.) They therefore did not restrict any conception of the golden ratio to the digits 618 which characterise phi, either in 0.618 (φ) or 1.618 (Φ) or 2.618 (Φ^2)

If, for example, the ratio 0.382:1 appears, for example, then the golden ratio simply *must* be present. If either of the ratios 0.236:1 or 0.764:1 appear, then chances are there is a golden ratio hiding somewhere in the text and further judicious investigation is required.

Nor is the golden ratio limited, by the way, to the Fibonacci set: 1, 1, 2, 3, 5, 8, 13, 21, 34, 55 and so on. Any sequence which is constructed, like the Fibonacci numbers, by adding the two previous terms to create the next one will show the golden ratio fairly rapidly.[199]

I have found phi and the Fibonacci numbers to be so psychologically restrictive when it comes to analysis (whether discussing it with mathematicians or non-mathematicians) that I have chosen to get away from them entirely by using *tau*, the name employed by Martin Ohm (and which is still in reasonably common use in Europe.)

τ or *tau* is approximately 0.618 to 1 and it is named for 'the cut' and refers to a concept first described in Greek geometry. The idea may well have been known long before, however it was Euclid who first mentioned a line cut into *akros kai mesos logos*—the division into 'mean and extreme ratio'.

Consider the various ways a line can be divided, starting with the simplest: the line is cut in half so that two equal segments are obtained:

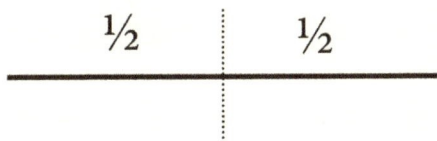

½   ½

Now because each half of the line is exactly the same, the pieces are in a 1:1 ratio with each other and in a 1:2 ratio with the whole line.

Now suppose the same line is cut one-third of the way along its length.

⅓   ⅔

This means one section is one-third of the whole and the other is two-thirds of the whole. The ratio of one section to the other is 1:2 and of each to the whole is 1:3 and 2:3 respectively.

The idea of *akros kai mesos logos* is to find the place on the line to make a cut so the ratio of the length of the first section to the second is exactly equal to the ratio of the length of the second section to the whole line.

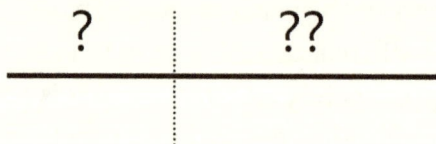

$$\underset{\textbf{?}}{\rule{3cm}{0.4pt}} \Big| \underset{\textbf{??}}{\rule{5cm}{0.4pt}}$$

This occurs at a position approximately 0.382 of the way along the line[200] which leaves 0.618 remaining. (0.382:0.618 :: 0.618:1) This is why 0.382 naturally defines a golden ratio just as much as 0.618 does.

Euclid didn't actually clearly define what he meant by *akros kai mesos logos*. This has important ramifications, because although the usual way of explaining the golden ratio is similar to the outline I have given above, it is not the only possible way of dividing a line into 'mean and extreme'.

There is an algebraic way of looking at the situation above. The solution which satisfies the following pair of equations is the golden ratio:

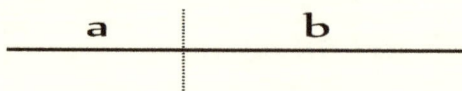

$$\underset{\textbf{a}}{\rule{3cm}{0.4pt}} \Big| \underset{\textbf{b}}{\rule{5cm}{0.4pt}}$$

$$a + b = 1 \text{ and } ^a/_b = b$$

However, medieval mathematicians of both East and West regularly conceived of the golden ratio as defined by a mean and *two* extremes. Abu Kamil, Al–Biruni, Morcedai Finzi, Gerard of Cremona, Adelard of Bath, Hermann of Carinthia and Campanus of Novara all spoke of a 'middle and two ends' or a 'mean and two extremes'.[201]

It seems to me that the most probable explanation for this is simply that they saw Euclid's definition as deriving from the geometry of a pentagram.

A straight edge and compass are the only tools needed to be able to create a 'division into mean and extreme' or 'mean and two extremes'. A 'golden rectangle' is a construction where the width is in a golden ratio to the length. This could be 6.18 cm and 10 cm, or 10 cm and 16.18 cm, or 10.5 cm and 17 cm: any size in fact which corresponds to the length being 1.618 times the width.

The most important geometrical figure to contain the golden ratio, however, is not a rectangle but a pentagram: the five sided star which can be drawn without lifting a pencil from the paper. This five sided star has been known since antiquity; it was once a symbol of Jerusalem and was variously called Solomon's Seal and the Endless Knot[202] as well as being the symbol of trawthe painted on Gawain's shield in *Sir Gawain and the Green Knight*.

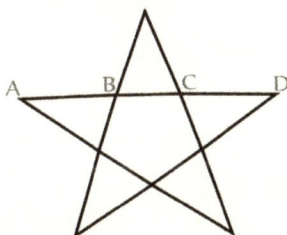

Consider the five–pointed star shown.
- The section AB is 0.618 of the section AC. (Strictly speaking, it is only *approximately* 0.618; please note that decimals quoted in this section are rounded to three significant figures.)
- It is also 0.618 of the section BD.
- AC is also 0.618 of the section AD.
- So is BD.
- In addition, CD is 0.618 of the section AC.
- BC is also 0.618 of the section BD.
- Furthermore BC is 0.618 of the section AC.

If we keep on looking for different examples of the golden ratio, we'd never stop. An infinity of golden ratio relationships can be found in this apparently simple diagram.

The pentagram suggests where the medieval notion of a mean and two extremes came from. In the diagram on the previous page, the line segment BC might be considered the 'mean' and the line segments AB and CD would be the two extremes. This gives a flexibility most modern reviewers overlook: B or C, and not just C, can denote a golden section or golden 'cut'.

The golden ratio aspects of the pentagram were well known long before medieval times. They are most frequently said to be a discovery of the Pythagoreans (indeed, very often attributed to Pythagoras himself in the sixth century BC), however as Roger Herz–Fischler points out, there is no evidence to support this claim. The first reference to the pentagram in connection with the Pythagoreans dates from the second century AD where it is mentioned simply as a symbol of health without any reference to the golden ratio whatsoever.

From the time Pythagoras lived to the time the pentagram was first connected with him is eight centuries—longer than the time between ourselves and the period when the *Pearl* poet was working. So if the pentagram was of such significance to the Pythagorean Brotherhood, surely some mention of their association with it would have occurred in the intervening time. The very fact it isn't connected with them in any context whatsoever suggests it is likely to be a late addition. Indeed, it's possible the widespread attribution of the discovery of the irrational properties of the pentagram to Pythagoras is completely erroneous.

Now if misconceptions about the golden ratio abound, then misconceptions about the pentagram are so numerous they make the former seem almost insignificant. Some studies of *Sir Gawain and the Green Knight* are predicated on the assumption the pentagram is a pagan symbol of great antiquity which the *Pearl* poet attempted to re–define away from witchcraft into a Christian ideal. Is the pentangle a pre–Christian symbol of life and 'power over the other world' which acquired Christian significance in this poem as is assumed by many editors and critics from Jessie Weston on? Or was there a long Christian tradition already in existence as the poet himself states?

Unfortunately the 'great antiquity' of association between the pentangle and pagan religion seems to have been back–dated from 1862 when Éliphas Levi asserted in *Ritual and Dogma of High Magic* that an inverted pentagram was a symbol of Baphomet which he associated with a goat's head and the devil. Prior to this, the pentagram was widely recognised as a symbol of God; the English magician, John Dee, possessed equipment marked with the pentagram as integral to the 'Seal of God'.[203]

The issue of the golden ratio is, however, slightly different to that of the pentagram. Long before any use of it can be definitely attributed to the Pythagorean Brotherhood, it was widely used as an aesthetic ideal in literature and architecture.

Greek literature contains references to poetry modelled on the dimensions of the human body and, in some cases, the poet's ongoing debate with himself about whether or not he is going to conform to the standard model. Isocrates, for instance, breaks off his story of Theseus because he has gone outside 'the due measures', however in a later oration on Agamemnon decides that, despite the criticism he knows he will receive for not keeping to these same due measures, he cannot omit some important details.[204]

This idea that 'Man is the measure of all things'[205] survived into the Renaissance when Leonardo da Vinci sketched his famous ideal man, based on the writings of the Roman architect, Vitruvius. Despite the colourful conspiracy woven by Dan Brown in *The Da Vinci Code*, Leonardo's fascination with the golden ratio has quite prosaic origins. He was engaged as an illustrator for a book on mathematics, *De Divina Proportione*—'The Divine Proportion'. Luca Pacioli, a Franciscan monk, is usually considered the author of that seminal work though it is sometimes claimed he simply plagiarised the notes of his mentor, the artist and mathematician, Piero della Francesca. Pacioli is often cited as believing that the golden ratio of the human body is indicated by the navel.[206]

This, however, as we shall see in the *Pearl* poet's work, does not seem to be an original observation.

De Divina Proportione is often credited as changing the nature of art—from its publication on, artists became conscious of the golden ratio as an aesthetic ideal for the first time since the decline of the Greek city states. Far from being a renaissance, however, *De Divina Proportione* was twilight's last gleaming. The golden ratio as a construction device in literature had all but disappeared and even in art, it was becoming severely restricted. The heyday of the golden ratio in medieval and dark age culture was over.

It's curious that despite Theodore Cook's extensive look at various aspects of Leonardo's work in *The Curves of Life*, he never mentions the term 'divine proportion' which was also used by Kepler. Indeed, Cook rarely used any existing terminology from the books he cited—his agenda clearly being to produce the impression there was a gap in the literature into which he allowed Schooling to introduce Mark Barr's term *phi*.

As a result, the widespread use of the golden ratio in medieval times has been obscured. In fact, it may even seem odd to suggest that the golden ratio enjoyed vast popularity in medieval times. However, many instances of it can be cited.

Ed Condren first noticed the use of the golden ratio in the poems of the *Pearl* manuscript. He mainly restricted himself to *phi* and considered that the poet had modelled his work on the geometry of the intricate carpet pages in designs such as the Book of Kells and the Lindisfarne Gospels.

Robert Stevick has shown that there are pages in the latter manuscript which have clearly been designed using the golden ratio. So too have the Muiredach Cross at Monasterboice, the leather cover of the Stonyhurst Gospels and the magnificent Hunterston Brooch. However, as the golden ratio is very simple to achieve geometrically, Stevick does not consider that there is any significance in its use, other than the usual aesthetic one.

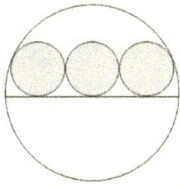

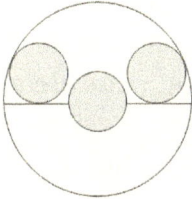

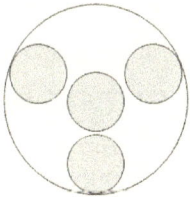

Nevertheless there is a carpet page in the Book of Durrow which I consider calls this view into question. There is an unusual geometric formation using the golden ratio which is not simple to achieve. It circumscribes 3 circles in a hemisphere: the diameter of the circles to the radius of the hemisphere is the golden ratio.

A modification of this design seems to be used as the basis of a design in the Book of Durrow. There is no significance in a single design motif, but the same page features no less than three other 'whirling circle' design elements which appear in my view to approximate the golden ratio: two of them within ½% in terms of area and one within 2%. Four different golden ratio designs on the same page, three of them very obscure, suggest that the golden ratio is not a mere construction device: it has defined meaning.

Stevick has shown that the tenth century poem, *Elene*, on the finding of the True Cross has a deep structure built around the number 618, the golden ratio of 1000.

Prior to the tenth century, David Howlett has found many examples of texts—religious, political and domestic—'sealed' by what he calls Biblical style. In an age when photocopying didn't exist and when scribal copiers were apt to be subject to human error, sophisticated numerical techniques existed for ensuring the accurate transmission of a manuscript.

A very simple example: *On the third day, when a mere eight knights were left, in the middle of the sky, there appeared a marvellous portent.*

The previous sentence is numerically sealed by virtue of the fact the third word is 'third', the eighth word is 'eight' and the middle syllable is 'mid'. It's not very tightly locked, but the general idea is there. Howlett has demonstrated that Biblical style was firmly established well before the Middles Ages and involved word, syllable and letter counts which enabled both error and fraud to be detected.

His comparison of parts of the Masoretic text of the Jewish Scriptures with corresponding sections of the Vulgate shows that Jerome, while unable to copy the exact word and letter count in translating from Hebrew to Latin, nonetheless tried to follow the overall style within the numerical literary design.[207] Howlett moreover notes golden ratio crafting used multiple times in the opening chapter of Genesis within the Masoretic text. Not surprisingly, there are also many multiples of 7 used within the text.

J. Smit Sibinga has noted a precisely similar technique in early Christian writing which he calls by the name 'numerical literary technique'. Maarten JJ Menken has analysed much of John's gospel and has concluded that the numerical style there is so superbly controlled that, in all the passages he has examined, no more than a single syllable in total has been lost.

Menken noted that any use of a number in the text itself is a clue to any numerical analysis of the passage: for instance, multiples of 38 are used to structure the scenes which feature the man who had been waiting by the Pool at the Sheep Gate for 38 years. He also noted that the Greek text of the Hymn to the Logos which famously opens John's gospel with 'In the beginning was the Word...' uses multiples of 17 in its construction. Smit Sibinga, like Menken, noted the repeated use of the golden ratio by different gospel writers, pointing out in particular that, in the Sermon on the Mount, the divine title 'Father' is used 17 times and it is distributed in such a way that the *Lord's Prayer* ('Our Father') is situated at its golden section.[208]

Smit Sibinga, Menken and Howlett all used arithmetic methods; Stevick and Condren favour a more geometrical approach.

With regard to the poems of the *Pearl* manuscript, I find it difficult to choose. On any occasion that I have thought that the evidence favours the arithmetic over the geometric, I have quickly found something to balance that view. And any time I've succumbed to the opposite view, proof duly emerged to suggest that the geometric was not favoured above the arithmetic, either. In the end, it is probably a mistake to suggest that a medieval poet would have thought of one approach as superior to another and used it only. Modern thinking may well divide geometry from arithmetic and both from literature but poets of the Middle Ages, educated in the quadrivium and trivium, naturally thought in integrated terms.

One of the important aspects of Howlett's work is his discovery of the use of the word 'mean' to mark the golden ratio. This, of course, makes perfect sense since Euclid's *akros kai mesos logos* was a 'division into extreme and <u>mean</u> ratio'. It's a play on words, suggesting that the *Pearl* poet's mathematical play on 'poynte' has a long and respectable history.

The first example of the use of the adjective 'golden' in conjunction with 'mean' at the golden ratio that Howlett cites is from a letter of John of Salisbury in 1165:

> *quoniam licet hanc <u>'auream mediocritatem'</u> quam*
> *prescribo seruare non nouerim aut non queam.*

While obviously this alludes to the philosophical golden mean of Horace, its positioning strongly suggests that it is a pun with mathematical undertones. John was a student of the renowned neo–Platonic philosopher, Thierry of Chartres, whose School is said to have influenced the Pythagorean design of many churches and cathedrals from this period on.

Also around this same time period, the first currently known use of the adjective 'golden' to indicate the golden section appeared in French poetry. Joan Helm noted the use of 'gold' as a significant marker word for golden ratio partitioning in the Arthurian romances of Chrétien de Troyes such as *Le Chevalier de la Charette* and *Cligés*.

Helm revived the suggestion of Urban Tigner Holmes that Chrétien de Troyes was a Jewish convert to Christianity. She amassed considerable evidence from Chrétien's works which show his understanding of Hebrew cosmology and measurement. She has furthermore proposed the use of the golden ratio in Chrétien's works was part of a twelfth century neo–Platonic revival centred at the court of Marie de Champagne, the daughter of Eleanor of Aquitaine and half–sister of Richard the Lionheart. She believes Chrétien, who undertook his work at Marie's behest and who indicated his indebtedness to her for suggesting both theme and treatment, hid the evidence of his enthusiasm for neo–Platonic ideas behind an all–but–impenetrable code of veiled allusion and arcane meaning. In her view, he hoped to escape the criticism of the church authorities in this way.

My question is: was it *really* hidden?

I don't think it was. In fact, it's my view absolutely no one was fooled. Contrary to Helm's view that the numerological aspects of the poem remained under wraps for over seven hundred years to avoid censure by the Church, I believe the whole of literate Europe was aware of them. The golden ratio wasn't something that could be swept under the carpet so no one would notice—it was simply too well known.

Quite apart from the design of the Lindisfarne gospels and such monuments as the Muiredach Cross, it was also used the cross of Durrow as well as in church construction, notable examples being the cathedral at Chartres,[209] Saint–Yved at Braine,[210] Notre Dame of Amiens, Saint–Ouen in Rouen and Saint–Urbain in Troyes[211] as well as Durham Cathedral.[212] It was also used in the ivory cover of the Golden Gospels of Echternach, in the Evangelist pages of the Macdurnan Gospels[213] and even in town planning for bastides or 'new towns' such as Grenade–sur–Garonne, Ste–Foy–le–Grande and the Zaehringer cities.[214]

Along with $\sqrt{3}$,[215] it was, according to Jan Tschichold, used for the area of the text inside elaborately bordered manuscript pages.[216] Most importantly for this analysis, it is found in the *Chanson de Roland*[217] and in the poetry of Cynewulf. This should be no surprise since the evangelists Matthew, Luke, John and Paul used it extensively while Roman poets such as Horace devised *Ars Poetica*[218] around it and Virgil featured it strongly in *The Georgics* and *The Aeneid*.[219] Robert Stevick has shown Cynewulf used it in *Elene* and mentions its use in *Christ II*. Chaucer was later to use it in both the *Parliament of Fowls* and *The Book of the Duchess*.[220] Various other fourteenth century poems such as *Divine Love* and *Winner and Waster* show evidence that Chaucer wasn't alone in appropriating it.

Silverstein points out it was also used in seal designs at the time. He also notes that, as far as the pentangle goes, in the sixteenth century, Pierius Valerianus and Cornelius a Lapide both associate it with the Five Wounds on the Cross (exactly as it is portrayed in the description of Gawain's shield). They apparently drew their knowledge from a fifth–century book of heraldry well–known in England and also refer to a Roman military unit called the *Propugnatores* who carried shields decorated with a pentagram.[221]

A brief examination of some of the different ways the golden section was used in four works of late medieval poetry will assist in assessing the *Pearl* poet's own work in the next chapter. Here I am looking, as Ed Condren did, for deep structural embedding,[222] and also in addition, like Joan Helm, for the themes and ideas highlighted at the golden section.

I am looking for a common pattern, similar to the one that can be found for the Kiss of Heaven and Earth in *Cotton Nero A.x* and Chrétien de Troyes' *Erec et Enide*. I am thus trying to find what the golden ratio meant to Chrétien de Troyes, Geoffrey Chaucer and the *Pearl* poet—not mathematically but rather metaphorically. Moreover I'm interested in what traditions they were working from. Chrétien and Chaucer were clearly operating out of a neo–Platonic framework; the *Pearl* poet seems to know the works of both his rivals intimately and also to know, if not the works of Cynewulf, the tradition that prompted the use of 618 as the defining number in *Elene*.

To find these traditions it's necessary to find the common elements. Recurring motifs which appear at the golden section of various poems are:

- gold
- peace
- truth
- love
- the Fibonacci numbers
- the 'triple triangle'

Gold, of course, should be expected as a key word in the light of Chrétien's exemplar at golden sections.

Indeed, yes, it is constantly found.

For instance, in Chaucer's dream vision *Parliament of Fowls*, the golden section (Line 267 of 699 lines or 0.382 of the way through the poem) refers to Venus and a golden thread,[223] thus not only harking back to courtly love but specifically alluding to Chrétien's *Cligés*. In that romance, Gawain's sister Sordamour sews a shirt with a blonde hair next to a gold thread in a totally bizarre test to check whether a male of the species (she doesn't have a particular man in mind, any will do!) can discern the difference between the two embroideries.

At Line 617 (where the distance to the end of the poem constitutes 0.236 or φ^3 of half the poem) Nature commands 'Peace!' in such a definitive way that the bickering birds fall obediently silent and the final movement of the poem begins. As mentioned earlier, any notable change at a 0.236 position suggests that a serious look for a hidden golden section should be undertaken. Particularly when, as in this instance, the last movement of the poem begins on a line whose digits are the hallmarks of the golden ratio: Line 618. That number is waving a banner, calling to us and telling us that there's another golden section.

The trouble is that there's nothing specific on Line 432. Surrounding it is a general sentiment which is appropriate: the protestation of a tercel that he will never be untrue should he win the hand of the formel. However, golden sections aren't hazy.[224] The allusion at a golden section may be obscure but the cut, if there is one, is sharp.

Line 432 of *Parliament of Fowls* is, of course, twice 216 lines. 216 is the pre-eminent Pythagorean number—it is the mystical number of re-incarnation, not only because it is $6 \times 6 \times 6$ (since 6 is a 'circular number' as 6×6 gives a number ending in 6) but because the 'divine Pythagoras' was said to be re-incarnated after 216 years.

Is Chaucer a neo-Pythagorean as well as a neo-Platonist? He explicitly mentions Pythagoras in *The Book of the Duchess* (immediately after Line 666 and then again on Line 1167.) Ed Condren notes that this might actually fulfil one of the stringent requirements of Peterson for analysing mathematical construction: that an author should state his intentional use of it.

A golden ratio exists in *The Book of the Duchess*, as Condren has shown, in the dialogues (which he separates carefully from the monologues of the man in black) which are partitioned according to a standard Fibonacci formulation of 89 to 144. Unfortunately, the monologues, which might have cleared up the matter of Chaucer's sympathies with regard to Pythagoras (the mention immediately after Line 666 is surely no coincidence!) are disputed: Condren sees the first and last monologue as 152 and 154 lines in length respectively and Shippey sees them as both 153.

If only it were some other number—*any* other number—than this one. Is Chaucer embracing 153 or trying to avoid it like the plague? That is *THE* question. And just to make matters even more ambiguous, a quick check of the golden ratio of the combined total of the monologues and dialogues reveals that it is 490. Is Chaucer asking forgiveness and, if so, what for?

In the end, *The Book of the Duchess* conceals more than it reveals about Chaucer's intentions and motives.

Returning to *Parliament of Fowls*, it should be noted it has an unusual asymmetric structure: the first part is 366 lines in length (corresponding to Victoria Rothschild's suggestion that there is a commemorative calendar for the leap year 1384 hidden in the text) and the second is 333 lines.[225] The total length of the poem is 699 lines.[226] The second section, as a multiple of 111, suggests it is a motif of the City of God—or perhaps, in the context of the poem, the Garden of Eden.

In many ways, *Parliament of Fowls* with its reference to 111 and a leap year corresponds to *Pearl* with its hidden 1111 concatenated lines and its not–so–hidden reference to 354 days in a synodic year. (Or is the allusion to 354 cubits from one corner of the Temple to the other?) Although the outworking is completely different in each case, the mathematical inspiration behind the two poems has an uncannily similar echo to it.

Pearl, however, is a tapestry of additional connections which goes far beyond Chaucer's ambiguous mathematics.[227]

It is a mistake to think that all poets who used the golden ratio as a construction device did so out of a neo–Platonic framework. Some of them were severely and critically anti–Platonic. Others of them, however, were unlikely to have had the first notion what Platonism was.

For instance, Advent Lyric Eight of the *Exeter Book* is 61 lines long. Like Cynewulf's *Elene* designed around the digits, 618, it seems to be advertising its construction. Indeed the 37th line, marking the golden section, mentions 'golden gate'. Suggesting that this is not random chance is the next Lyric in the sequence which also mentions the 'golden gate', this time on Line 44. Moreover Line 44 marks the golden section of that poem's 73 lines. The golden gate in this lyric not only refers to the coming of Christ but to Ezekiel's vision of the City of God. It is a metaphor for the virgin birth and for heaven touching earth.

The *Exeter Book* dates from around the late tenth century, thus indicating that Chrétien was by no means the first to use either 'gold' at a golden section or play with the mathematics of the New Jerusalem. It precedes him by at least two centuries. What Chrétien does do, however, is fuse the mathematics of the heavenly city with neo–Platonism and with occult ideas from alchemy. Moreover implicit in his manner of use of the golden ratio was his flaunting of it as a token of courtly love. Thus not only *gold* and *peace*, but *love—true love*—is a likely battleground for a mathematical joust.

Not unexpectedly therefore, treuthe, trwe luf, trawthe and luf appear prominently at golden sections in various poems.

A fine example of this is *Divine Love*,[228] said to be written about 1375. In it, there is a prologue of 10 lines, followed by verses of 4, 4, 4, 4, 4, **2**, 4, 4 and 4 lines. There are 44 lines in total, but if the prologue is disregarded, there are 34 lines. These 34 lines are consistently made of four–line verses, except in one instance. 34 is a Fibonacci number with golden sections of 21 and 13. From the sudden change of regular four–line verses to that singular two–line verse, it might be expected that the golden section falls on either Line 31 or 32.

And indeed it does.

Lines 23 and 31 are both significant in terms of the golden ratio, one being the location of the 0.382 position and the other of the 0.618 position. (Because there is going to be so many statements like this coming up, please consider from this point on that, simply for ease of reference and in order to avoid too many awkward repetitious phrasings, the term *tau* refers to the location of the 'upper' golden section 0.618 and that *paz* refers to the location of the 'lower' golden section at the 0.382 position.)[229] However, it should be pointed out that if the line count is taken backwards instead of forwards, then Lines 24 and 32 are the significant ones.

Thus Lines 23, 24, 31 and 32 all need to be carefully examined. Line 23 mentions *treuthe*, Line 24 *trewe love*, and both Lines 31 and 32 feature *love*.

Ross Arthur has quoted numerous passages[230] in which treuthe and trawthe often had a doubled sense: they are not simply truth, integrity and faith but The Way, the Truth and the Life: God himself. This use of truth as a token of the golden ratio goes back at least as far as Cynewulf's use of it to construct the story of the finding of the True Cross in *Elene*. Once again this points to poetic design which pre-dates the peculiarly Pythagorean 'mathematics of truth' in Chrétien's romances by some two centuries and lends further credence to the poet's implicit avowal in *Sir Gawain and the Green Knight* (Lines 664–665) of a learned tradition which links trawthe to the pentagram and is Christian but apparently not Platonic in nature.

Divine Love is not the only poem to use a Fibonacci sequence.

While in some ways it's hardly fair to choose *Summer Sunday* as a point of comparison to the work of the *Pearl* poet because, as I previously mentioned, I so strongly suspect it to be his work anyway, the use of the Fibonacci numbers is so consistent in it that it's an irresistible choice for a brief overview.

As translated by John Gardner,[231] *Summer Sunday* is comprised of:

- six verses of 13 lines
- followed by a single verse of 8 lines
- then a single verse of 13 lines

- another verse of 8 lines
- two final verses of 13 lines

Thus there are 9 verses of 13 lines and 2 of 8 lines, a total of 133 lines. (As previously mentioned there is a dourad at Line 42 where Lady Fortune begins to spin her wheel.) Each 13–line verse is composed of 8 long lines and a 'bob and wheel' of 5 lines. The 5, 8, 13 construction recalls the Fibonacci numbers and is suggestive of a golden ratio design.

This is borne out by inspecting the golden section of the poem and noting that it is situated right in the middle of the first 8–line verse at Line 81. It would appear that it is positioned in the middle, rather than at the beginning or end of the verse, so that the number of lines from the golden section to the start of the next 8–line verse totals 17 and that the number of lines to the end of the poem is 51 $(3 \times 17$ lines.)[232] Between the middles of the two 8–line verses are 21 lines, another Fibonacci number. The middle of the second 8–line verse is $1/\tau^3$ from the end of the poem, suggesting that a very careful golden ratio design, coupled with a 'Wheel of Fate' construction was in place.[233]

Lastly, before we turn to the golden ratio in the work of the *Pearl* poet, it is important to consider the question of multiple constraints.

It's no trouble to produce perfectly–cleaved cuts with *gold* or *peace* or *trawthe* at the golden section if that's the only mathematical device underlying the text. But when a Wheel of Fate and a golden ratio, not to mention a triangular circle and over a thousand 'zeroes' are planned, then unless the numbers are really amenable, it's going to be a tussle. The poet had to decide which is the most important because multiple mathematical constraints pull different ways at the text.

I'm sure, for instance, that his contemporary Chaucer would have preferred *Parliament of Fowls* to have been 100 stanzas of 7 lines each. The fact that it is 699 lines clues us in to what was most important to him in the design: the leap–year calendar and the geometry of Eden.

Similarly the author of *Sir Orfeo*, which is generally dated around the mid–fourteenth century, would probably have preferred 605 lines (a multiple of 11) instead of 604. The fact that there is a dourad anyway, a change of fate at Line 192 where 'the queen was suddenly snatched away'[234] doesn't suggest that the manuscript has lost a line. It suggests that there was something far more important to the poet than a multiple of 22 or 11. 604 lines can be divided into a golden section of 231 lines and 373 lines. Here we have the clue: the author obviously wanted a golden section of exactly 231 lines.

Now I don't believe he wanted it because 231 happens to be a multiple of 11 (or even because a compass rose[235] of 22 points has 231 linking lines). I believe he chose it because it is triply triangular number—an arithmetic parallel to the geometric triple triangle (or Endless Knot, pentacle or pentagram) made so famous in *Sir Gawain and the Green Knight.*

Triangular numbers belong to the sequence starting 1, 3, 6, 10, 15, 21, 28, 36, 45, 55; their name comes from the fact they can be drawn as a triangle of dots. 231 is the 21st triangular number, which means it is formed from a triangle of 21 rows. It is called triply triangular because 21 is also a triangular number—the 6th—and 6 is the 3rd triangular number.[236] As triply triangular, 231 is said to be symbolic of the Trinity.

Thus 231, being an arithmetic allusion to the triple triangle so thoroughly explicated in *Sir Gawain and the Green Knight,* is an oblique reference to *treuthe*—not only a synonym for God—but, in fact, one of the dominant themes of the poem. Orfeo defeats the king of the faeries by appealing to *treuthe* in all its multifarious connotations of truth, faithfulness and integrity; later, on his successful return from the land of the dead with Heurodis, he disguises himself in order to test the *treuthe* of his steward.

However, a poet doesn't really need 231 to allude to *treuthe* or *trawthe,* as we have already seen with *Divine Love* and will further see in *Sir Gawain and the Green Knight.* The golden ratio itself is sufficient. So, what is it about 231 that made the poet choose it as his dominant constraint over a multiple of 22?

Its factors. That's what determined his choice.

The aliquot factors of 231 are 1, 3, 7, 11, 21, 33 and 77. They add up to 153.

Say no more.

How could a serious Christian author of the medieval age, a poet who knows that words and numbers are an indivisible unity, even consider writing about a descent into hell without somehow featuring the number 153?

It's inconceivable.

It is Jesus' own choice for a multiple of 17.

Which is why *St. Erkenwald*, which in its own subtle way features a harrowing of hell, has both 231 and 153 hidden in its design. It's one of the few things that is subtle about *St. Erkenwald*, so as we conclude this introduction and at last begin our examination of the use of the golden section in the works of the *Pearl* poet, let's start there.

Chapter 6

Trawthe and Trinity: the ratio 0.382:1

The number 231 is part of the hidden architecture of the miracle poem, *St. Erkenwald*. By examining its triply triangular nature in that context, we will be able to scrutinise the *Pearl* poet's golden ratio workmanship—which is much more complex, fulsome and multi-faceted than that of his contemporaries—in his simplest formulation.

Despite the consensus of opinion which at the present time appears to favour the views of Larry Benson, it is apparent to me that *St. Erkenwald* was written by the *Pearl* poet. The mathematics really leaves no doubt at all about it.

There is no question in my mind that the driving passion of the *Pearl* poet is shared by the writer of *St. Erkenwald*. It is not obvious on the surface, but it is certainly obvious in the mathematical tokens and in the references, all the more powerful because of their extreme subtlety, to the legends of Gregory the Great. The poet's style of arithmetic architecture, his 'numerical signature', the relationship between his mathematics and his themes, are so similar in both Cotton Nero *A.x* and *St. Erkenwald* that it forcibly suggests all five poems were written by the same person. He may have mellowed by the time he finished *Pearl* and *Patience* but the same blunt agenda remains.

Yet what does 'style' or 'numerical signature' mean with regard to structural design and arithmetic architecture? In the *Pearl* poet's case it means coming time and again to the same mathematical tokens and, through them, alluding to the same concerns.

He likes the number 5: it appears in the fivefold fives describing the qualities of Gawain's shield and person, it appears in the total number of saints referred to in *Sir Gawain and the Green Knight*, it appears in the shape of the pentagram with its golden ratio sub-structure. He is far from the only poet of the time with a predilection for the golden ratio but his intensity of use surpasses his contemporaries. It seems to have a particular meaning for him that goes beyond simple *trawthe* into a deeply personal philosophy.

Because that philosophy has its roots in the numerical literary design of the New Testament, he also likes the number seventeen[237]— as does the author of *Death and Liffe*. He likes circularity as we've seen and he likes triangular numbers. With the possible exception of the zeroes, nothing he does is original, but the way he puts it all together is astonishingly creative.

The simplest way of looking at the poet's concerns, his 'style' and his creative genius is to examine what themes he positions on his golden sections.

St. Erkenwald was probably written in 1386 as St. Paul's was being lavishly re-built and London was celebrating and commemorating its 'first' prelate, Erkenwald.[238] It seems to have been dashed off in a hurry.

Two feast days were proclaimed by Bishop Braybroke for the saint (30th April and 14th November) and general veneration was widespread. The poem appears to have been written expressly for one of these notable occasions.

St. Erkenwald has a number of phrases peculiar to the *Pearl* cycle, but that is not the major reason I think they share a common authorship. Not many poets are obsessed with the number 17; in fact, there are some who will do anything to avoid it. Condren may well be right about the optional nature of stanzas 17 and 70 in *Parliament of Fowls*. After all, the *Pearl* poet has lines that can function as 'zero' all over his work. However, if Condren is right, then Chaucer may well have been playing a very curious game.

My reasons for believing *St. Erkenwald* was written by the

Pearl poet is that, first, they both share a distinct predilection for 17 and, secondly, *St. Erkenwald* duplicates the same overarching mathematical format as *Sir Gawain and the Green Knight.* It also has a lot in common with the mathematical themes of the *Pearl* manuscript as a whole.

However the deep parallels with *Sir Gawain and the Green Knight* are most notable. They are equally why I think it's a rush job. Given the extent of numeric artistry within *Cotton Nero A.x*—the dodecahedron described by Condren as well as the elaborate scheme of pentagram, wheel and Cross within *Sir Gawain and the Green Knight*—it is obvious the *Pearl* poet had a substantial treasure chest of mathematics to draw on.

He used a Fibonacci–like sequence in *Purity*[239] and a 'sum of perfection' in *Patience* which we have only partially examined. As we've seen, he experimented with zeroes, sprinkling them liberally through the text. So the very idea of duplication in design suggests a lack of planning time for his poetic architecture and a decision to use a template ready at hand.

Alternatively, perhaps *St. Erkenwald* was a trial run for *Sir Gawain and the Green Knight.*[240] However, it doesn't strike me that way. It strikes me as a rush job. It must be remembered that a splendid alliterative poem doesn't have to take long to compose. Witness the testimony of Egil Skallagrimson.

When faced with death having fallen into the hands of his enemy Eirik Bloodaxe of York, he was advised by one of his friends to forget all about sleeping that night but to spend the time composing a drápa of twenty stanzas in Eirik's praise. Egil took the advice and the result was so good that it put Eirik in an invidious position. The Vikings valued a good poem far more than gold and Eirik realised that even though he dearly wanted to execute Egil he'd have to let him live if the poem was to become part of the wider skaldic repertoire.[241]

Still it must be admitted that even if an alliterative format is more forgiving for fast composition, *St. Erkenwald* isn't anywhere near the standard of *Pearl*. It does not display the same opulence of device, depth of allusion or nuance of wordplay. It isn't a masterpiece of suspense, fun and delicate irony like *Sir Gawain and the Green Knight*, either. Certainly it may be argued that it's much shorter and its scope is therefore considerably less. However the even shorter length of *Summer Sunday* does not seem to have been a constraint on its lyrical jewel–like construction.

The theme of *St. Erkenwald* deals with the question of the fate of the righteous man who had never had an opportunity to know Christ. The subject proper begins when the corpse of a pagan judge is discovered. The dead man comes from a time stated in convoluted fashion to be 354 years before the birth of Jesus. His flesh, fresh and sweet–smelling, is still on his bones when a sarcophagus is unearthed by workmen, hard at work re–building St. Paul's sometime in the seventh century AD.

The inspiration for this miraculous preservation of the dead judge's body is clearly the legend of Harold Godwinson, the last of England's Anglo–Saxon kings. The tale of the survival of Harold after Hastings was extremely popular in the twelfth and thirteenth centuries around Chester where, so the story ran, he came after many wanderings. Taking up residence as a hermit first in a cave at Dover, then in Wales, he was led at last by an angel to Chester and the church of St. John's where he ended his days. 'According to an apparently reliable Welsh annal compiled very soon after the event, in 1332 the body of Harold, clad in leather hose, golden spurs and crown, was found in the church of St. John smelling as sweetly as on the day it was buried, allegedly over two hundred years earlier.'[242] Indeed, even into the twenty–first century, the cowled one–eyed ghost of Harold is said to haunt St. John's,[243] prowling along a lost underground passage from the river Dee up through the ruins of the church.

St. Erkenwald consists of a Prologue of 32 lines, Part I of 144 lines, Part II of 176 lines: a total of 352 lines.

The table of the parts is therefore:

	Prologue	Part I	Part II	Total
	32	144	176	352
Ratio	2 :	9 :	11 :	22

Parts I and II are 320 lines and, if the poem was written in 1386 as suggested, then this may allude to the more common tradition, extant beyond Chester, that Harold died 320 years before at Hastings in 1066.

The Prologue and Part I together are the same size as Part II and thus the end of Part I effectively divides the poem perfectly in two. Musically this division indicates an octave.

The appearance of 22 in the total indicates that the manuscript can be divided evenly by 22, while the 11 indicates the possibility of a hidden hexatonic scale.[244] As in *Pearl*, *Sir Gawain and the Green Knight* and *Summer Sunday*, it seems the poet has structured his text with the concept of the Wheel of Fate in mind.

The dourad occurs on Line 112. It does indeed correspond to a changepoint. Erkenwald arrives back in London, having first been properly introduced only four lines earlier. He dashes in on a white horse, returning from a visit to an abbey in Essex.[245]

With only a few exceptions, the poem is composed of alliterative stanzas of four lines. All but three of its 88 stanzas follow this simple pattern. The 30th stanza however has five lines, the 38th stanza also has five lines and the 42nd stanza has two lines. These structural anomalies occur at lines 121, 154 and 167/168.

Ignoring the Prologue for the moment, the *paz* position of Parts I and II of *St. Erkenwald* is Line 154, just where the second structural anomaly begins.

Now, considering the Prologue and Part I: the *paz* is at Line 67 where the tomb is sealed off from its surrounds and the *tau* is situated at Line 109, only three lines different from the Wheel of Fortune position (Line 112) and one line different from the first major mention of Erkenwald (Line 108).

An important issue should be pointed out at this juncture: I'm using an irrational surd for my calculation while the poet may have used a fraction (and not necessarily a decimal one). The fact that he knew the decimal system doesn't mean to say he preferred it. Thus a difference of a few lines may simply reflect his procedural error and it is possible that he actually intended the Wheel of Fortune and this particular golden section to coincide.[246]

However, as I've indicated previously, I think we're dealing with a very meticulous man and, even granted it was a rush job with a greater possibility of error appearing, I don't believe the difference between the golden ratio and Wheel of Fate placements is necessarily an error of procedure.

If we turn our attention back to *Pearl* for a short while, let us recall that the line mentioning the wheel of the Divine Potter (line 386) is the *paz* of 1010. This brings up not only another multiple of 101, but also links a golden section to a dourad. Moreover, the very next line—Line 1011—mentions beryl, a stone according to Britton Harwood symbolic of the Five Wounds on the Cross.[247] These Five Wounds are connected to the pentangle (and thus the golden ratio)[248] in *Sir Gawain and the Green Knight*.

This reference to beryl is another linkage between 490, the token of forgiveness, to 101, the token of the Music of the Spheres. The entire assembly of identical mathematical ideas in *Pearl* and *Sir Gawain and the Green Knight* is so clearly paralleled by those in *St. Erkenwald* that it's a pity beryl isn't referred to on the previous line of *Pearl*: it seems to negate the notion of the meticulous artist I've just been promoting.

The reason for this single line discrepancy is, however, comparatively easy to discover. There is another constraint in play. If we return to Line 858 (which we noted previously as a complete wheel of *wyrd* and also noted is so suggestively 354 lines from the end of *Pearl*), we see that it is 152 lines away from the 1010 position whose *paz* refers to the Divine Potter. I suspect the poet felt just as I do: *if only it were 153 lines*. So near and yet so far.

Unless we count inclusively, no amount of numerical massaging will produce a result where the lines exactly coincide. There was clearly only one thing to do: the poet moved the 'golden' allusion from Line 1010 to Line 1011, in the process also managing to avoid a multiple of 10.[249]

Line	Link	to Line	Token
1010	× 0.382	386	Music of spheres is linked to a destiny determined by Divine Potter through trawthe
1011	−153	858	Complete cycle of destiny is linked to trawthe through the number of resurrection

The poet actually has a number of these tightly woven[250] paired lines, most of them next to each other, but sometimes as much as five lines apart. What at first sight is a mathematical fudge to allow for multiple constraints may well have become an essential part of the poet's strategy. Two, after all, is the 'number of witness' in Hebrew thought. Under the Law, the testimony of two or more witnesses was required for a judgment. Importantly, these witnesses to *trawthe* did not have to be human. Moses not only called on 'heaven' and 'earth' as witnesses to the covenant between God and Israel, God later called on them to testify that Israel had broken the covenant. Joshua called on stones as witnesses. The two letters 'aleph' and 'taw' often remain untranslated in the Hebrew Bible since they serve as witnesses to the truth.[251]

Extending the idea of two witnesses to numbers is perfectly logical in these circumstances, particularly when it seems that Paul of Tarsus did exactly this in his letter to the Corinthians. Indeed, as we shall soon see, the design numbers that Paul uses at one significant point in that epistle are exactly the same as those used by the *Pearl* poet in *Sir Gawain and the Green Knight*.

So all this means that I suspect that the paired lines (either Lines 108 and 109 or Lines 108 and 112) in *St. Erkenwald* are specifically intended. They're by no means as tightly woven as the paired lines of *Pearl*, but this only suggests once more that *St. Erkenwald* was composed under considerable time pressures.

Now while it's correct to say, as Condren did in his analysis of the *Pearl* Manuscript, that it is unlikely the poet was aware of the Fibonacci numbers or knew of the connection between them and the golden ratio,[252] it should not be overlooked that, as mentioned previously, *any* similar additive sequence will tend towards *tau* in the ratio of its higher terms. The only constraint on the choice of initial numbers is that at least one of the two is non–zero. Apart from that, any numbers at all can be picked. Having said all this, it should be also noted that in *St. Erkenwald* the design numbers actually coincide more often than not with the Fibonacci sequence.

The anomalous two–line 42nd stanza is 8 lines from the end of Part I. 13 lines back is the anomalous five–line 38th stanza which corresponds to the *paz* position of Parts I and II. 34 lines further back is the anomalous five–line 30th stanza and 55 lines further back is the *paz* position of the first half of the text, the Prologue and Part I.

Thus, out of the Fibonacci series of 1, 2, 3, 5, 8, 13, 21, 34 and 55, the poet has—either by good luck or good management—chosen to highlight 2, 3, 5, 8, 13, 34 and 55 in various ways. In fact, he manages to highlight another one as well in the size of Part I which is 144 lines long. 144 is the 12th Fibonacci number, it is the square of 12 and its square root is 12 and thus it is reminiscent of the dodecahedral symbolism underpinning *Pearl*.

The only missing numbers at this point from the first dozen of the Fibonacci sequence are 89 and 21.

There are two formal text divisions and another three irregularities making five highlighted partitioning points—the end of the Prologue, the three anomalous stanzas and the end of Part I. These occur at lines 33, 121, 154, 168 and 176. Poor cousins to the illuminated initials of *Sir Gawain and the Green Knight* which, according to Aya Peard, symbolise the Five Wounds on the Cross,[253] they nonetheless seem to reflect dimly the 'fives' which dominate the Arming Sequence and the description of Gawain's shield.

Of the five partitioning points, four of them are divisible by 11. These are 33, 121, 154 and 176. This pattern is reminiscent of the way the illuminated initials in *Sir Gawain and the Green Knight* are positioned—four regular and one irregular, the latter representing the fifth wound, the spearthrust in the side.

Even more interesting, however, is what happens if we consider these partitioning points in a different way. Some of these breaks seem to fall naturally on the first line of a section, some of them on the last. However, let's re–configure so that, *with one exception*, they fall on the last line.

Line	To next anomaly or partition point	To end
32	89	320
121	32	**231**
153	13	199
167	9	187
176	–	176

As we can see from the first line of the table, the missing 89 from the Fibonacci sequence has turned up. And so has 21, although in a heavily disguised fashion: as the 21st triangular number 231. 153, the 17th triangular number, has also appeared. Because of their triangular format, both these numbers hark back to the tetraktys, the ultimate symbol of Pythagorean truth.

The triply triangular 231 is reminiscent of the 'triple triangle' found in *Sir Gawain and the Green Knight*. The 'triple triangle' is another name for the pentangle, so we have here an arithmetic echo of a geometric form, both of which connote the Trinity.

For the poet, the triple triangle of Sir Gawain's shield betokens *trawthe*, a word meaning truth, faithfulness and integrity, a word frequently used as a synonym for God himself. Moreover, *trawthe* perfectly reflects the Hebrew concept of the pentangle: truth.

Even at this point, it's clear there's a remarkable congruity between the thinking of both the *Pearl* poet and the writer of *St. Erkenwald*: extraordinarily similar ideas, if not exactly the same numbers, are in constant association. There is a consistent theological pattern in the approach to mathematical construction. The pattern explodes into the light when we examine what the poet chooses to feature at his golden sections.

In *St. Erkenwald* with its ten featured Fibonacci numbers, the poet was clearly playing golden ratio games throughout his text. It seems strange then that, at first sight, the *tau* position of the text should look so unpromising:

During the reign of the royal monarch who ruled us then,
the bold Breton, Sir Belin—Sir Brennius was his brother—
many an angry insult was offered between them,
various the violent wars while their vengeance lasted,
I was appointed principal judge here of the pagan law.
While the sepulchred man spoke, there sprang from the people
Not a word in all the world, not awoke one sound,
As still as stones, they all stood and listened

Lines 212–219

translated by Brian Stone

Line 217 looks so pedestrian that it is almost impossible to conceive of it as the raison d'etre of the poem's format, the heart of the inspiration for the design of the entire poem. Looking back a few lines to see what the words of the sepulchred man actually are, we find they are so detailed and so definitive it is surprising they have been overlooked.

We don't know the name of the sepulchred man himself, but we are offered the information that he was a judge during the time of the various violent vicissitudes of Belin and Brennus. Not just any Belin, thank you. The Breton. Just to be sure we know exactly which Belin the Breton, we are given the additional detail that Brennus was his brother. Now why should either of these two be associated with the golden ratio? And who are they anyway?

In several excruciating and tortuous lines which date the foundation of London to 1136 BC, the pagan judge reveals that he held court during 354 BC. Brian Stone makes the point that the poet's sources actually put the reign of Belin at 354 years *after* Christ and therefore claims that the poet clearly erred at this point. I'm not sure we can be so certain about the sources of this man of 'fructous intelligence'. I think he fully intended to refer to the two Celtic warlords who really did pre–date Christ by some three centuries and who, once they started to fight together rather than against each other, even conquered Rome. I also think this dating reference was so important to the poet that he felt compelled to make his audience suffer some bad poetry in the interests of thumping his point across.

A contemporary reader, on encountering this reference, might have thought of the passage in Wace where King Arthur is reminded by Hoel, one of his barons, of the Sibyl's prophecy that three kings of Britain should conquer Rome. Brennus was the first, so Hoel maintains, Constantine the second and obviously, it is inferred, Arthur will be the third. Arthur himself claims to be a near kinsman to Brennus, Belinus his brother, Constantine and Maximilian—each of whom he proclaims was a Master of Rome.

Brennus was also mentioned in the history of Geoffrey of Monmouth.

The date given by the poet is so close to the total line numbering of 352 that we might wonder if it's simply coincidence or whether there is some connection. 354 is, as I've mentioned several times previously, the number of days in the old synodic year: 29.5 days in a lunar month and 12 months in a solar year combine to give 354 days in the synodic year. 354 is thus a symbolic reference to the time of the sun and the moon. Not separately, but together: a combination that suggests yet again a union of heaven and earth.

It is also very likely, as also mentioned previously, that it is a mathematical metaphor for the Temple within the City of God. Consequently, it is my belief that 354 was the poet's defining number for this poem, as 490 was his defining number for *Sir Gawain and the Green Knight*.

Unfortunately, however, 354 is not divisible by 22.

But 352 is.

Moreover, 352 is a multiple of both 176 and 12. Both 176, 22 and 12 are part of the English system of length (12 inches in a foot, 22 yards in a chain and 1760 yards in a mile) and these are a reflection of a spiritual understanding about God the geometer and the measurement of the cosmos. Thus, in my view, practical constraints forced the poet to abandon his preferred defining number in favour of one that could reflect, not the sun and the moon, but Hebrew measurement[254] plus a hexatonic musical scale as well as provide the opportunity for the traditional changepoint of the Wheel of Fate.

This same mathematical problem is solved slightly differently in *Purity*: there the poet uses 352 lines for the story of Noah and 353 lines for the dourad of 1111 lines, thus using a pair of lines to unite part of a terrestrial measurement system to a heavenly one.

Still, within *St. Erkenwald*, the poet wanted everyone to know that his defining number was really 354.[255] Moreover, he wanted everyone to know that, although it might be associated with the synodic year in popular thinking, he wanted it to evoke the story of Belin and Brennus. In doing so, he was providing a vital clue with regard to his aim.

A sideways look at the Arming Sequence of *Sir Gawain and the Green Knight*—the famous passage where Gawain's shield is described in sumptuous detail and with extravagant explication—hints at more about Belin, Brennus and this vital clue.

> *Forþy þe pentangel nwe*
> *He ber in schelde and cote*
> **As tulk of tale most trwe**
> *And gentylest knyȝt of lote*
> *Fyrst he watz funden fautlez in his fyue wyttez.*
> *And efte fayled neuer þe freke in his fyue fyngres.*
> *And all his afyaunce vpon folde watz in þe fyue woundez*
> *Þat Cryst kaȝt on þe croys, as þe Crede tellez.*
>
> Lines 636 – 643
> *Sir Gawain and the Green Knight*

The *paz* position of the second 'half' of the *Pearl* Manuscript falls on Line 638 and is in fact a 'double *paz*' speaking of Gawain, the man of truth—a truth symbolised as the preceding two lines state in the pentangle emblazoned on his shield and surcoat. Blanch and

Wasserman see the gloss on the pentangle as 'overlong', its very length calling into question the poet's powers of judgment. Tolkien agrees, suggesting that his taste is shared by his contemporaries and that it is elaborate and pedantic because the description proved too difficult for the author's skill with alliterative verse.[256] However, alliteration and rhythm are not the poet's primary concern. While the gloss is indeed long, it is long for a very precise and particular purpose.[257]

And that purpose is mathematical and metaphorical.

Falling on exactly the same spot (Line 638) is the *paz* of the two stanzas which feature a detailed exposition of why the poet has chosen the ultimate golden ratio emblem—the pentagram—as the perfect heraldic device for Gawain's shield with its emblem of truth.

Gawain, we're told, is a 'tulk' of truth. Is the choice of 'tulk', *man*, merely for alliterative effect? Or is there more to it? A word derived from Old Norse, its original sense was that of 'interpreter'.

Gawain, the *interpreter* of truth, the man who speaks on behalf of truth, perhaps hints that this is the heart of the rhetorical argument of Lady Truth.

We need to play close attention as we attempt to interpret Gawain's shield with its pentangle on the outside.

And its face of the holy Virgin painted inside.

Blanch and Wasserman see the shield as an attempt by the poet to 'gild the lily'—that is, to use the pentangle on the outside as a spell against magic should the protection of the Virgin fail. For this idea to hold, I suggest that it would be much more acceptable at every level for the Virgin to be on the outside of the shield and the pentangle inside. It's unnecessary to invoke magic as the reason for the pentangle. In fact, it doesn't make sense. Gawain's failure of *trawthe* is a result of succumbing to the lure of magic and trusting in the girdle, rather than God.

Nonetheless, I take the point that there may be more to here than meets the eye. Is this is a hint to look for a mathematical construction similar to *Lilja*? In fact, is Gawain's devotion to Mary is really the point of this detail at all? The poet's view of Mary as 'Queen of Heaven' must be viewed through the lens of his theology in *Pearl* where the Margery–maiden is one of thousands of queens in heaven. What other allusion could be here in Gawain's shield apart from Mary as protector?

The description with its emphasis on red and gold suggests it is a deliberate evocation of Pridwen, the famous shield of King Arthur which had a picture of the Blessed Virgin engraved inside it in red gold outlines. Pridwen means *white form* or *fair face*, similar in meaning to its Welsh counterpart, Wynebgwrthucher, *face of evening*.

Pridwen however was not only the name of Arthur's shield, it was also the name of his ship. This odd duplication of name is often explained as a simple confusion of words. Sometimes however, following Michael Wild's suggestion who notes that *fair face* is a very odd name for a ship,[258] it is seen as a reference to the only known example of a shield being used as a ship—that of the Celtic warlord, Brennus, who used shields as rafts during his campaign in Greece.

Brennus.

Again.

In two different poems at a golden section.

Surely this is not coincidence.

Both *St. Erkenwald* and *Sir Gawain and the Green Knight*, one overtly and one very subtly, refer to the wars of Brennus and in particular to combat in Greece. What historical point is the poet trying to make and why is it so important to him that he structures two poems around it? Why does he choose the golden ratio in both cases to highlight it?

It is my contention that these two allusions both hark back to the attack of the Gallic tribes on the temple at Delphi.[259]

In fact, in case his audience missed the fact that Gawain's shield[260] was modeled on Arthur's Pridwen and that it was a veiled allusion to the sack of Apollo's sanctuary, the poet has built in another reference from an entirely different angle. The fiveways fives which describe both the pentagram and Gawain himself hark back to Plutarch's famous fiveways fives in his discourse, *On the E at Delphi*. These fiveways five emerge in a long comment on the Platonic solids and Plato's view of the senses, matter and the universe.

By Plutarch's own testimony, no one had any idea what the

E on the navelstone at Delphi meant. It was so ancient by the first century when Nero[261] inquired as to its meaning that Plutarch could only make several educated guesses. Amongst other more religious possibilities, he suggested it might simply mean 'if'.

There was more than one E at Delphi: there were apparently a gold one, a brass one and a wooden one, the gold one being designated by the priests of the oracle as that of Livia, wife of Caesar. Coins have been found at Delphi with E on them, as have amulets. It looks like an ancient logo, one that was still sufficiently well-known in the twelfth century for it to be appropriated to mark the Pythagorean mystical division of *The Knight of the Cart*: the golden section. Joan Helm, in detailing all the golden sectionings of Chrétien's work marked by 'gold', was baffled by the lack of gold at *the* golden section of the entire manuscript: a mysterious illuminated E is there instead.

There is no doubt: Chrétien's work is not merely neo-Platonic and not merely neo-Pythagorean. I believe its enigmatic E refers to the navelstone[262] at Delphi and to worship of Apollo. Although the work of Plutarch allegedly was not available in medieval Christendom, supposedly only being translated in the fifteenth century, I feel there is sufficient circumstantial evidence to proceed as if both Chrétien and the *Pearl* poet were fully aware of the basic subject matter of *On the E at Delphi*, if not familiar with the text itself.

Furthermore, it is my contention that, without a basic understanding of Chrétien's mathematics and the philosophy behind it, it is impossible to appreciate the *Pearl* poet's choices or his motives.[263] To suggest that he loathes the work of Chrétien is vastly underrating his passion. He so detested the French notion of courtly love and was so intent on attacking 'all the golden gods the Gauls still call on'[264] that Chrétien's themes roll up again and again, even in the most unlikely places.

This is true for the poet's use of the golden ratio, it is true in the way he re-iterates the motifs of Cynewulf while attacking those of Chrétien, it is true in his choices of theme and it is true in his strategy.

In the late fourteenth century, the symbolic rose of courtly love had been replaced by a daisy: a marguerite (or heliotrope) daisy, to be precise. But a marguerite is not just a daisy, it is also a pearl, indeed the eponymous margery–pearl of *Pearl* itself. The petals, the 'white leaves' of the daisy, were in fact compared to pearls.[265] This is yet another reason I believe it's futile looking for a John or Hugh Massey who had a daughter Margery or Margaret who died in infancy: the pun of pearl is directed primarily towards the works of Guillaume de Machaut[266] and Jean de Meun. It may also impart an occasional dark glance towards the marguerites and heliotropes of a certain Geoffrey Chaucer.[267] It's the *Romance of the Rose* that the poet was attacking: explicit in Lines 269–272 of *Pearl* is the thought that the *pearl is made out of a rose.*

We have already seen that the mathematical metaphors of *Pearl* cluster on the poem's internal theme of the heavenly city. It is therefore a natural choice to have the archetypal Child as the Pearl maiden. After all, *unless you become as little children, you will not enter the kingdom of heaven.*

Editors such as WRJ Barron have long noted the similarity between the works of Chrétien and the *Pearl* poet but the depth of antipathy has gone unremarked. *Sir Gawain and the Green Knight* has been recognised as raising a glittering standard to declare itself an opponent of courtly love but *Pearl* has not been seen in the same light. However it is just as much a reaction to Chrétien's influence.

Perhaps it was Lancelot swooning and swooning and swooning some more at the sight of one of Guinevere's golden hairs on a comb as well as Alexander, the father of Cligés, receiving a shirt with a golden hair[268] embroidered on it (the very clues that led Joan Helm to discover the connection to the golden ratio) which influenced the *Pearl* poet's choice of champion. Instead of Guinevere of the golden hair, he chose fair–headed Gawain whose very name, according to Roger Loomis, comes from 'Gwallt Avwyn' meaning *hair like reins* or 'Gwallt Advwyn' meaning *fair–haired.*[269] Gawain's name moreover reflects the first syllable of Guinevere, even when both undergo serious changes of spelling.

The influence of Chrétien is to be found even in the poet's clichés. He makes the point in the opening of *Sir Gawain and the Green*

Knight that knighthood came to Britain through the bravery and gallantry of the Trojan hero, Brutus. He's far from alone in parading this particular 'historical' detail for his audience. It is almost a pre-occupation of English poetry. Perhaps the reason it is so common has far less to do with conservatism in poetic convention than it has with Chrétien's provocative comment that chivalry made a natural stop when it got to France. 'Gawain's adventures take place in a framework that originates in the fall of Troy, the founding of Rome, and racial migrations through Italy and France that lead up to Brutus's founding of Britain. By tracing the genealogy of Arthurian knighthood, the poet implicitly refutes Chrétien's assertion in *Cligés* that chivalry passed only from Greece and Rome to France (the motif of the *translatio studii*.)'[270]

The opening of *Cligés* both parallels and contrasts with the beginning of *Sir Gawain and the Green Knight*, which is quite ambiguous in the way it is worded. There is difficulty in working out exactly what the reference to the 'treason of Troye' means and who the traitor is.

Is this because there's a pun involved? Is it because there is an allusion to both Troy and Troyes? Is the traitor from Troy or from Troyes? Should the question not be, 'Is the traitor Aeneas or Antenor?' but rather 'Is the traitor Aeneas or Chrétien?'

Moreover, if there is a pun on Troy and Troyes, was the *Pearl* poet the first to use it? Where exactly have the sunbeams that pick up all the 'story of Troye' and all the 'Romaunce of the Rose' in Chaucer's *Book of the Duchess* actually been dallying?[271]

A passage from *Cligés* may reveal why the *Pearl* poet may have felt there was much to be said for a traitor from Troyes. Fenice, the Empress of Constantinople, on confessing her love for her husband's nephew, reveals that, despite all appearances to the contrary, her marriage has never been consummated. The Emperor has drunk a magic potion which gives him the illusion of nightly sexual satisfaction. Fenice of course wants to be with her true love and has hatched a plan. She does not, however, propose following the example of Tristan and Iseult, but suggests instead faking her own death. She quotes St.

Paul's advice to Timothy in advocating this course of action, pointing out that the apostle taught that those who found it too hard to remain chaste should be careful to arrange their affairs so that no one finds out what they're doing. Be discreet and so avoid criticism and slander. (*Cligés, 5306–5312*)

Chrétien thus not only claimed brazenly in *Cligés* that France alone had inherited a legacy of chivalry from the classical past, but had also twisted the words of Paul regarding sexual morality so that the emphasis was not on remaining chaste but on avoiding a bad reputation so that the church is not impugned. Not only that, his magic–dabbling Scripture–perverting heroine was called Fenice, a name clearly derived from 'phoenix,' the mythical bird which lived 500 or 540 years and renewed itself in the fires of its own funeral pyre. The phoenix, because of this resurrection through the flames, was understood throughout the Middle Ages as a type of Christ.

If this wasn't enough of an attack on a traditional Christian motif, to add mathematical insult to theological injury, Fenice, lying in a death–like coma, is attacked by three doctors from Salerno who see through her ruse and begin to assault her on Line 5940—which at 540 × 11 is a multiple of one of the traditional lifetimes of the phoenix— and set her on a fiery pyre on Line 5994—or 54 × 111.

Did Chrétien know that 111 was a number associated with the City of God and the reign of justice and peace? According to Joan Helm's analysis of *Erec et Enide*, yes, he most certainly did.[272]

So whether Chrétien actually intended to be offensive or whether he was simply intent on re–defining various Jewish and Christian touchstones in order to frame a neo–Gnostic and Pythagorean background is probably immaterial. Offence was taken in his own time and moreover was still being taken over two hundred years later.[273]

There's a distinct anti–French sentiment pervading the entire text of the *Pearl* manuscript—sometimes subtly, as in the opening refutation of the *translatio studii* in *Sir Gawain and the Green Knight*, and sometimes overtly as in the swift knife–swipe in *Purity* about 'all the golden gods the Gauls still call on'. While Chrétien is certainly a target, the *Pearl* poet's agenda may be far wider.

The four poems of Cotton Nero *A.x* were almost certainly all composed during the Western Schism and the era of the French anti-popes. In 1378, the election of Urban VI was claimed to be invalid by the cardinals of France who instead put forward Clement VII as the legitimate heir to the See of Peter. Clement took up residency in Avignon, France and for the next forty years until the death of Benedict XIII, the scandal of two—and sometimes three—popes rocked western Christendom.

In locking his sights on Chrétien de Troyes, perhaps the *Pearl* poet was attempting to preserve a delicate balance: by attacking the schismatics so indirectly, he could avoid the danger of fostering disrespect for the papacy and the wider hierarchy of the church.

The heresies of Chrétien, although two centuries prior, may have provided the poet with a perfect backdrop to criticise the Avignon line of popes.

The area around Troyes had long been a centre of neo-Platonism. Over a millennium previously—as far back as the third century in fact—the druids of Gaul were reputed to be stalwart followers of Pythagoras. In possible confirmation of this assertion, considerable numbers of hollow bronze dodecahedrons have been found in the region dating back to that time. Little is known about the dodecahedrons. However, it has been suggested that the druids allowed smoke to pass through them in order to produce omens for health prognostication.[274]

It is from this same period of time—the third century—that most of the earliest writings about Pythagoras emanate. The sage of Samos had died nearly 800 years previously and fragmentary stories existed prior to the third century, however if there were any complete histories about him or his Brotherhood, none have passed down to us.

In the fourth century one of the great promoters of Pythagoras emerged. After a season of turmoil following the death of his uncle, the emperor Constantine, Julian eventually assumed one of the thrones in a divided empire. Although he had been brought up as a Christian,

Julian had turned to Pythagorean theurgy under the influence of Maximus of Ephesus, a student of Iamblichus. Resolving to replace Christianity with neo–Platonic paganism, Julian nonetheless appears to have been of the opinion that Christianity was, in fact, the people's choice. In his view, much more than an imperial decree was needed to get rid of it. What was lacking in his estimation was a moral philosophy which could rival the Christian ethical code. Sending to the oracle of Delphi for advice, he promoted the writings of Iamblichus, particularly *On the Pythagorean Life*, as the ethical basis of a return to paganism.

The Pythia reputedly delivered the Delphic Oracle's final message:

Tell to the king that the carven hall is fallen in decay;
Apollo has no chapel left, no prophesying bay,
No talking spring. The stream is dry that had so much to say. [275]

Julian the Apostate's short reign did much to ensure the fame of *On the Pythagorean Life*. One of its catechetical questions referred to the oracle at Delphi and to Python Apollo, the tutelary deity after whom Pythagoras was named.

Question: What is the oracle of Delphi?

Answer: The tetraktys. It is also the harmony in which the Sirens sing.

It was the tetraktys or tetract which was the ultimate mystical symbol of Pythagoreans: 10 dots arranged in a triangle.

As mentioned previously, it was regarded as Manifest Deity, the source of nature, the Number of Numbers, the Meaning of Meaning, the creative principle, the fundamental Truth of the universe, the heart of the Logos.

So, hidden in this enigmatic catechetical answer is a claim that Pythagoras (as either Pythian Apollo or Hyperborean Apollo) is the creator of the cosmos and, since the Sirens were supposed to sing one note each of the musical scale, its sustainer as well via the Music of the Spheres.[276]

Chrétien's use of the Delphic E at the golden ratio had these overtones: it was a claim that the tetraktys is the source of ultimate truth. It implicitly denied the centrality of the True Cross. Such a claim may have been acceptable at the provincial court of Troyes but certainly it didn't go down too well in the wilds of fourteenth century Cheshire.

The notion that 'courtly love—the love that a knight might have for a lady of higher rank than himself—leads to spiritual ennoblement'[277] was evidently anathema to the *Pearl* poet. Yet if his anti–Chrétien overtones are as thorough–going as I am suggesting[278] and, if he knew about the meaning of the E, it could be expected that he would have featured 'e' in a prominent way. The difficulty, of course, would have been in coming up with a concept opposite to that of Chrétien. How does a poet go about giving a letter *negative* prominence?

As long ago as 1921, Oliver Farrar Emerson noted that the imperfect metrical lines of *Pearl* and *Sir Gawain and the Green Knight* could be amended and made regular by simply adding a final unstressed 'e'. There are several dozen cases of this allegedly missing 'e' and despite Emerson's suggestions that the poem's dialect could indeed comfortably accommodate the original form, I incline to a different suspicion. The first three lines[279] which require an amendment of an additional 'e' to make them regular are 17, 51 and 68. All of these line numbers are multiples of 17.

Pythagorean mysticism is not just a religion of numbers, it had long been tied to theurgy and a tradition of magic. This made it particularly welcome in third century France where it was said to be strongly linked to druidism and where, in the twelfth century, it had a major resurgence. There was only one number the Pythagoreans loathed, only one number which according to Plutarch, high priest at Delphi in the first century—the same writer who penned *On the E at Delphi*—they considered to be an '*atrocity*' or '*abomination*'. This was the number 17. They called it the antiphraxis, variously translated *precaution, interposition, disjunction, obstruction* or *barrier*.[280]

Plutarch relates that their abhorrence had to do with the dismemberment of the Egyptian god, Osiris. However this explanation seems remarkably odd to me. I'm not inclined to totally discount it but I have to consider the possibility that Plutarch was dissembling. Pythagoras was after all named after the very shrine where Plutarch was high priest: the temple of Python Apollo at Delphi. Why they would be concerned about the sacred rites of an Egyptian god is baffling, since the Pythagoreans were noted for *not* being religious—that is, for not worshipping a multiplicity of gods, but sticking to one. Like the Christians and the Jews before them, they were considered to be atheists because they did not worship a pantheon of deities but were devoted to one god. And 'The One,' in the case of the Pythagoreans was not Osiris, but Apollo.

Most of what we know about the legend of Isis and Osiris actually comes from Plutarch (which, if we look for a modern equivalent is essentially the same as relying on the Dalai Lama for the main beliefs of the Shinto religion—expert on religion as he is, should we really rely on him to explain the subtleties of any belief system other than his own?) As a result, I believe we should be very wary indeed of Plutarch's possibly disingenuous comments about 17. This is especially the case since that number had already firmly established itself as *the* favourite amongst the writers of a small sect of Judaism, variously known as followers of The Way or as Christians.

It was a number peculiarly favoured by the gospel and epistle writers. As already mentioned, the structure of the Sermon on the Mount in the Gospel of Matthew is built around 17 references to the Father, which are positioned to create a golden section at the Lord's Prayer. Matthew also uses the term 'Christ' 17 times. Luke in his description of the descent of the Holy Spirit in the Book of Acts mentions people of 17 ethnically distinct groups who heard Peter's sermon at Pentecost. Paul in Philippians makes 17 references to joy,[281] mentions 17 qualities desirable for an elder in his first letter to Timothy and 17 names in his personal remarks at the end of his second letter. He uses *Sophia*, 'wisdom', 17 times in 1 Corinthians while in Romans 8: 35–39, he lists 17 things that cannot separate a Christian from the love of God. There are 17 uses of the word, *nikao*, 'conquer', in Revelation; in his gospel, John qualifies *zoe*, 'life', 17 times with the adjective 'eternal' and uses the name Simon for Peter

17 times as well as the word *semeion*, 'sign' or 'miracle' 17 times, the same number of times as he uses *eido*, 'know the truth', in his first epistle. In the 'Bread Section' of Mark's gospel from Mark 6:30–8:10, there are 17 references to 'bread'. In his roll-call of faith, the writer of the Hebrews mentions 17 patriarchs.[282] John, in the opening Hymn to the Logos at the beginning of his gospel, structures the Greek text around multiples of 17—in fact his first sentence has 17 words.

These examples are by no means exhaustive.[283]

Such inescapably obvious numerical favouritism would not have eluded any mathematically-minded poet.

Thus Condren's discovery that Stanza 17 of *Parliament of Fowls* is structurally dispensable suggests Chaucer was side-lining those verses in much the same way as the early Pythagoreans had wanted to remove the 'abomination' and that his Christian faith was struggling with neo-Platonism at the time. The fact that the stanza in question contains a *non sequitur* in which the presence of Venus is suddenly invoked simply confuses the issue.

As we turn back to the *Pearl* manuscript, realising that Plutarch was apparently not translated into Latin until the fifteenth century, we have to consider that his comments on both seventeen[284] and the navelstone of Delphi were widely known or else that the *Pearl* poet read Greek.

We also have to consider that means the poet was not dependent on the Latin Vulgate but was able to read the New Testament in the original language. If we assumed that he did in fact know Greek fluently, this would have a singular advantage: we could then assume it was unnecessary for him to know the Hindu–Arabic decimal system, since Greek numbers, although not written in a decimal fashion, were *spoken* that way.[285] Hierocles in his fifth century commentary on the *Hieros Logos* of Pythagoras clearly sets out the archetypal decimal system:

Now the finite interval of number is ten, for he who would reckon after then comes back to one, two, three, until by adding the second decad he makes twenty, by adding the third in like manner he makes thirty, and so on by tens until he comes to a hundred. After a hundred he comes back again to one, two, three and thus the interval of ten always repeated will amount to an infinity.

Nonetheless, this does not solve all difficulties since Greek mathematics, like the Roman, had no zero, while the *Pearl* poet unquestionably makes use of the concept of 'nought'.

Pythagorean and anti–Pythagorean allusions abound in the *Pearl* poet's work. In *St. Erkenwald*, the golden section at Line 217 adjoins Line 216, the mystical Pythagorean number of re–incarnation. Line 216 explicitly refers to 'the pagan law' and it may well be a spiritual law (that of re–incarnation) that the poet has in mind as much as social law.

St. Erkenwald is pointedly NOT about re–incarnation. Like *Sir Gawain and the Green Knight*, it features the pre–eminent number symbolic of resurrection: the number 153. And moreover it links 153 to the triple triangle, a token of trawthe and thus of God.

153 is the 'resurrection number' for a mathematical reason, a theological one and also a Scriptural one.

Dicit eis Jesus: Afferte de piscibus, quos prendidistis nunc. Ascendit Simon Petrus et traxit rete in terram, plenum magnis piscibus centum quinquaginta tribus. Et cum tanti essent, non est scissum rete. Dicit eis Jesus: Venite, prandete. Et nemo audebat discumbentium interrogare eum: Tu quis es? scientes, quia Dominus est. Et venit Jesus, et accipit panem, et dat eis, et piscem similiter. Hoc jam tertio manifestatus est Jesus discipulis suis cum resurrexisset a mortuis.
John 21:10–14 (Vulgate)[286]

There are so many puns and allusions in this gospel passage that it's difficult to know where to begin. There's a reference to Archimedes and $\sqrt{3}$ wrapped up in a joke about measuring or counting the number of fish.[287] There's a possible allusion to the navelstone at Delphi in the description of the unbroken net.[288] There may be a dig at Pythagoras in the description of Jesus' invitation to breakfast.[289] There's a clue as to why John thought the number of the fish so significant that it was worth actually recording it.[290]

None of these have anything much to do with the theological reason usually advanced for the appearance of 153 in medieval poetry: Augustine's comment that, as the 17th triangular number, it represented 10 lines of natural truth and 7 lines of spiritual truth or the gifts of the old and new covenant—the ten commandments and the sevenfold Spirit.

There's barely the faintest whiff of evidence that the *Pearl* poet took any notice whatsoever of Augustine.[291] In fact, he seems to have ignored Augustine's view on this matter entirely. And, because he did, it's worth taking a fresh look at the mathematics.

The mathematical reason for 153 as a symbol of 'resurrection' is threefold:

> (1) The digits of 153 are 1, 5 and 3. Taking the numerical literary hint in 'this was now the *third time* Jesus appeared,' each of the digits is cubed, that is multiplied by itself three times. This yields $1^3 = 1$, $5^3 = 125$ and $3^3 = 27$. $1 + 125 + 27$ adds up to 153. Thus 153 is the first number which can be 'resurrected' from its own skeleton by cubing its digits.[292]

> (2) $1 \times 2 \times 3 \times 4 \times 5 + 1 \times 2 \times 3 \times 4 + 1 \times 2 \times 3 + 1 \times 2 + 1 = 153$ (or in other words, if the world were made up of 5 different elements, as the ancient believed, then perhaps 153 represented to them the sum of all possibilities)

> (3) 153, as we have already seen is the sum of the aliquot numbers (that is, the whole number factors not including the number itself) of the 'triply triangular' number 231. Moreover 231 can be constructed as the sum of 153 and 78, both of these in turn triangular numbers. The use

and significance of triangular numbers from the earliest Christian times should not be underrated, either in a positive or negative sense. The most famous number in Christianity—666, the Number of the Beast—is the 36th triangular number. Whether it was ever used by the Pythagoreans as a symbol of re-incarnation[293] is conjectural, but it fits the typology in every respect.

There's not much evidence the poet used any of the mathematical sub-structure of 153 except perhaps for its relationship to 231, but there are a few hints he knew about the puns in the Greek text of John's gospel. It's impossible to be sure, of course, but it seems likely he did. 5 out of the 12 illustrations accompanying the text have a thematic unity of fish: big fish and little fish which, at least in the case of Jonah, suggests a resurrection motif since the 'sign of Jonah'— three days in the belly of the fish—had been equated with death and resurrection since the early Christian era.[294] Indeed, in Line 231 of *Patience*, Jonah is cast into the sea, the storm's raging surge ends and the 'sign of Jonah' begins.

In *Sir Gawain and the Green Knight*, on the other hand, Line 372 (corresponding to 231 × 17 lines into the whole manuscript) shows the subtlest of resurrection motifs: Arthur advises Gawain to take care and deal *only one* blow to the Green Knight. This counsel reflects the warning given to Pwyll by Arawn in The First Branch of *The Mabinogi*: Pwyll, prince of Dyfed, goes hunting a white stag which turns out to belong to Arawn, lord of the underworld. Pwyll offers to pay compensation and Arawn asks that Pwyll take his place in the underworld to fight his enemy, Hafgan. He advises Pwyll to take extreme care to strike Hafgan only once—if Hafgan is struck twice, he will return from death.

Now in *Sir Gawain and the Green Knight*, the resurrection number 153 is tied to the golden ratio in a distinctive way. As previously mentioned, the *paz* of the two stanzas which describe Gawain's shield and the *paz* of the second half of the manuscript (*Patience* and *Sir Gawain and the Green Knight*) both fall on Line 638. However the pre-eminent theological reference to the golden ratio is four lines away on Line 642/643 where the Five Wounds on the Cross are given prominence. The Five Wounds are in fact a doubly subtle way of referring to 'gold' without actually mentioning the word. The feastday of the Five Wounds on the Cross had been in the church liturgical calendar since the sixth century, but it was given new prominence in the fourteenth century when the 'Golden Mass' was indulgenced in either 1362 by Innocent VI or in 1334 by John XXII; during its celebration five candles were always lighted and it was popularly held that if anyone should say or hear it on five consecutive days he should never suffer the pains of hell fire.[295] This is the first hidden reference to gold. It also hides two more references to fivefold matters, making a total of seven in all.

The second hidden reference is more explicitly mathematical: the Five Wounds on the Cross was a common medieval motif often symbolised as stigmata which were themselves symbolised as the star–shape within an apple core or as a five–petalled blossom. The Five Wounds were given special political significance in 1139 when Afonso Henriques was about to face the Moors in battle at Ourique in Portugal. Like Constantine eight hundred years previously, a miraculous portent gave Afonso the assurance of victory. This vision of the Five Wounds on the Cross is still recalled even today in the Portuguese flag which retains a dim memory of the old golden ratio symbolism in its unusual $^2/_5$ and $^3/_5$ colour divisions.

Given the connection between the Five Wounds on the Cross and the golden ratio, why didn't the poet mention the Wounds on Line 638 instead of Line 642/643? It could be that the poet was using 0.619 (or $^{13}/_{21}$) as his approximation for the golden ratio—that would account for the precise discrepancy in the case of the 'half' manuscript (but not the two stanzas). There is a hint that this might be so. The illuminated initial which appears so unexpectedly at the start of these stanzas is on Line 619. However, this may simply reflect thinking that place a cut for the golden section immediately *after* Line 618, rather than our modern way of thinking which would be to place it *on* Line 618.[296] My suspicion remains that the poet was using 0.618 and that the reason the Five Wounds on the Cross is mentioned on Line 643 rather than Line 638 is a simple one: 643 is 153 lines beyond 490.[297]

Resurrection and mercy referenced by a Golden Mass which thematically speaks of assured resurrection through assured forgiveness: a perfect metaphor.

Moreover this line is also adjacent to Line 644 which is 17 × 111 lines from the end of the poem.

Thus the golden ratio is linked to the defining number of the manuscript through 153: trawthe is linked to forgiveness through the power of the resurrection. It is also linked to the Kingdom of Heaven through the arithmetic metaphor of 111.

A similar construction is used in *Pearl*: 231 lines from the end of the poem is Line 981, adjacent to Line 980 (or 2 × 490). In these lines, the Dreamer or Jeweller appropriately begins to move towards the celestial Jerusalem which he has just spied after its descent from heaven. 153 lines forward is the description of the *fifth* wound of the Lamb—the spear thrust in the side. This is again a picture of the heavenly Jerusalem since all three mentions of the wounded Lamb occur in the book of revelation in the description of the throneroom. One example is: *I saw a Lamb, looking as if it had been slain, standing in the center of the throne, encircled by the four living creatures and the elders. He had seven horns and seven eyes, which are the seven spirits of God sent out into all the earth.* (Revelation 5:6 NIV)

To complete the design, 153 lines back is the end of Stanza 69 and Folio 49[r] and, thus, the beginning of the section which deals with the vision of Revelation and the descent of the heavenly Jerusalem.

These numbers, so important to *Pearl*, *Sir Gawain and the Green Knight* and *St. Erkenwald*—153, 231, 490 and the golden ratio— are found together in the numerical literary structure of chapter 9 of Paul's first letter to the Christians at Corinth.[298] In addition, the number 170 (17 × 10) is also found there. All these numbers cluster on verses 14 and 15 in one way or another.

In the same way, the Lord has commanded that those who preach the gospel should receive their living from the gospel. But I have not used any of these rights. And I am not writing this in the hope that you will do such things for me. I would rather die than have anyone deprive me of this boast.

1 Corinthians 9:14–15 (NIV)

This idea, that a man should receive a fitting fee for his work, is found on Line 1358 of *Sir Gawain and the Green Knight*, 4913 (or 17 × 17 × 17) lines into the manuscript. It may be nothing more than a coincidence that the *Pearl* poet shares these particular structural numbers with Paul and connects them to a due reward for work done. However, if it is not simply an accident, then perhaps we have unearthed a hint about the poet himself and his occupation. Was he one of those priests so abjectly poor that he worked for a living? While many churchmen lived in comparative luxury at the time, at the other end of the scale there were also priests so hard-pressed that they kept taverns or ran farms to make ends meet.

Certainly, Jordan de Holme in 1369 took leave of absence from his parish and during the time was given permission to let his church out to farm.[299] It is unnecessary to suggest, as has often been done, that the poet can only have been a widely-travelled member of the court of Edward III or Richard II because of his accurate description of rich, fashionable armour, chantry architectural styles and detailed hunting techniques. The intelligent mind seeks stimulation and records tiny snippets of conversation: the poet's keen observations could have come from pilgrims, parishioners or princes. This is not to say that the *Pearl* poet was not a member of the aristocracy or of the court, simply that it is not a foregone conclusion.

Regardless of his personal circumstances, however, the letter to Corinth was on the poet's mind. Lines 458–462 in *Pearl* constitute the *paz* of Lines 1200–1212. And it is here that several important clues to the poet's thinking congregate:

457   *Al arn we membrez of Jesu Kryst:*
           As heued and arme and legg and naule
           Temen to hys body ful trwe and tryste,
460   *A longande lym to þe Mayster of myste.*

Line 458 is a clear reference to 1 Corinthians 12:14 and yet the poet refers to the navel which

(a) he must have known wasn't in the epistle, and

(b) he doesn't need for the purposes of alliteration, and

(c) is clearly stated as belonging to God, here given the unusual title of 'Master of spiritual mysteries,' thus echoing the way the prophet Daniel spoke of God in Daniel 2:28–29 and also Nebuchadnezzar, king of Babylon, later used to identify God in Daniel 2:47.

Granted that arms and legs aren't mentioned in the Corinthian epistle either, but feet and hands are. It seems strange that 'hand' is omitted when it so aptly alliterates. It seems even stranger that 'navel' is preferred over the far more obvious choice of 'eye' which is referred to several times by Paul.

Yet Line 458 clearly belongs to a *paz* position and so the navel—widely accepted in the Renaissance as the golden section of the body—is appropriately 'marking' the spot. This suggests that, long before the Renaissance, the navel was widely accepted as marking one of the golden ratios of the body.

However the description of God as 'Master of spiritual mysteries' suggests that this is not simply a reference to the human body but, yet again, to the navelstone at Delphi and to the religious mysteries of Pythian Apollo.

The oracle at Delphi was situated high in the cliffs above the Bay of Corinth and Paul's letter to the Corinthians clearly makes continual allusion to its influence on the thinking of the Christians who lived in the city across the waters.[300] Indeed the *Pearl* poet was much less subtle than Paul. The limbs of clay or metal offered as prayers or thanksgiving to Apollo the Healer were almost certainly in Paul's thoughts as he wrote his letter but in the poet's hands they are less a reference to the church community than to ownership by the Master of spiritual mysteries.

As we now embark on a swift survey of the lines corresponding to various golden sections, note how often the poet alludes to either a tearing down of the power of Delphi or to a corrective view about it. He continually and constantly had in his sights the destruction of the Pythagorean eidolon in his own time—the ideals of courtly love epitomised in the romances originally emanating from twelfth century Troyes.

To begin the survey, let's review the *tau* positions, that is, the lines which are 0.618 of the way through the various poems.

TAU

- The *tau* of *Pearl*, Line 749, is traditionally seen as referring to the *Roman de la Rose* and thus is another corrective to the ideas of courtly love. (An extremely curious feature in the stanza closest to this *tau* is its disruption of the usual concatenation. Line 757, just after a line divisible by 7, does not quite follow the usual first–line/last–line repetition. To add to its unusual character, it is marked in the manuscript by what appears to be a fist or a hand. This may be an early example of a 'manicule', a symbol very similar to the modern 'fistnote'.[301] These fist–like marks are pictorial clenched hands with pointing fingers which focus the reader's attention towards an important selection of the text. The choice of a fist suggests a highlighting mechanism and seems like a clue in itself. Perhaps it is a reference to a 'fistmeile', a measurement derived from closing the hand into a fist and raising the thumb.[302] Still its significance is by no means apparent but it is clear there is no coincidence in finding all these three highlights clustered so closely together.[303] However their overall meaning is elusive.)

- The *tau* of *Pearl* when considered as only 1200

lines is Line 742 which explicitly mentions 'pes,' *peace*. In fact it specifically states that the pearl is a sign of peace. This suggests that to view *Pearl* as the rhetorical set-piece for one of the Four Daughters of God—Lady Peace—is not only appropriate but is explicitly stated in the text.

- The *tau* of *Purity*, Line 1120, refers to an unspecified white stone. It may be a pearl or it may be the navelstone at Delphi which, while not white, was sometimes considered so from its marble covering.
- The *tau* of *Patience*, Line 328, is regarded as a reference to Psalm 85:10—*Mercy and Truth meet together, Peace and Justice kiss each other.* (Vulgate Psalm 84, 11: *Misericordia et veritas obviaverunt sibi; justitia et pax osculatæ sunt.*) This not only mentions the Four Daughters of God but brings together two common motifs for golden sections: *trawthe* and peace.

Patience has a prologue of 60 lines—the *tau* of the prologue is the much discussed lines 37 and 38, which 'fettle' poverty and patience, the first and last of the Beatitudes which not only herald the coming of the Kingdom of Heaven, but as a whole might be summed up by Psalm 85: 10 with its reference to mercy and truth, justice and peace.

- The *tau* of the body of *Patience* (that is, minus the prologue) is Line 351: Jonah, having been delivered from the whale, sets off for Nineveh.
- The *tau* of the whole manuscript falls on Line 206 of *Sir Gawain and the Green Knight* which refers to the holly bob in the giant's hand. Here again is an allusion to peace.
- The *tau* of *Sir Gawain and the Green Knight*, Line 1564, and the *tau* of the last half of the manuscript, Line 1361 of *Sir Gawain and the*

Green Knight, both refer to animals at bay.

That seems far from coincidental but how it fits in the overall scheme is not immediately clear. Does this allude to Gawain being held at bay by Lady Bertilak? The illustration which shows her approaching Gawain in his bed is accompanied by the couplet: '*Mi minde is mukel on on þat wil me noȝt amende/Sum time was trewe as ston and fro schame couþe hir defende.*'

Line 822 of *Pearl* uses this same simile 'trwe as ston' to describe the Lamb of God. It should be pointed out that 'ston' also occurs on Line 2063 of *Sir Gawain and the Green Knight* which happens to be the 5618th line of the entire manuscript. Given the emphases on 5 and on 618, the defining digits of the golden ratio, 5618 does seem far too suggestive a number for a *mere* stone, particularly when the 5618 × τ^4 = 820, just two lines different from *Pearl*'s mention of 'trwe as ston'.

There's clearly a significant allusion here to a matter not entirely clear. Like the workers in *St. Erkenwald* who look on 'still as stone' as the judge describes his tenure during the reign of Belin and Brennus, Gawain stands 'stylle as þe ston' in Line 2293, which (counting inclusively) is 153 lines from the 6000th line of the entire manuscript.

- The *tau* of these 6000 lines is 153 lines into *Sir Gawain and the Green Knight* and, while there does not appear to be anything of significance at that particular point, nonetheless the line numbering with its repetition of 153 suggests there might be a significant allusion missed or overlooked.

It should also be noted that the combined length of *Patience* and *Sir Gawain and the Green Knight*—the second 'half' of the manuscript—is 153 lines more than 1111 ÷ τ^2, which may be meant to evoke (and yet go beyond) a similar structural formation in Chrétien's *Erec et Enide*.[304]

This pre–occupation with 153 indicates a focus on resurrection—a concept which was anathema to the Gnostics who believed that the body was a prison for the spirit. They had no hope in the resurrection of the dead because, for them, bliss involved freedom from the constraints and corruption of matter, not a return to it.

The overall impression is that 'trwe as ston' has a specific cultural or religious meaning, which the poet utilised in such a way that it could operate as an oblique allusion to Delphi, its navelstone and its claim to be the source of ultimate truth. Moreover, it could also operate as a counterclaim that only the Lamb of God really has the right to assume such a title. He alone was Trawthe and Trinity.

PAZ

Let us turn now to a consideration of the *paz* positions, that is, 0.382 of the way through the various poems.

- The *paz* of *Pearl* has already been considered in the discussion of the Master of Spiritual Mysteries, however it should be noted that, at 462 lines, it is twice 231 and thus has embedded in it a triply triangular number with 'resurrection' factors.

- The *paz* of *Sir Gawain and the Green Knight*, Line 1170, refers to men skilled at the hunt pulling down animals which had escaped the archers at a receiving line.

- The *paz* of *Patience*, Line 202, sees the sailors questioning Jonah.

- The *paz* of the prologue of *Patience* is Line 23 which mentions a clean heart and links mercy and peace, directly linking to the *tau* of the poem and the *tau* of the prologue.

- The *paz* of the body of *Patience* (minus the prologue) is Line 240, translated as 'and acknowledged him alone to be God and truly no other.'[305]

- The *paz* of *Purity*, Line 692, finds God speaking frankly about the value of sexual love—in marital context, of course. This not only reflects a concern with courtly love on the one hand, but on the other with the neo–Platonic ideas of the Cathars of Languedoc who, influenced by Gnosticism, held that matter (and therefore the body) was entirely evil and was created by Rex Mundi, the 'king of the world'. Their view was that the true God was the creator of love, order and peace and, as a pure spirit, was unsullied by mortal flesh. Marriage was to be avoided in Catharism.

- The *paz* of the first half of the manuscript is Line 1155 of *Pearl*, the desire of the jeweler to be with his pearl beyond the water.

- The *paz* of the second half of the manuscript is Line 638 of *Sir Gawain and the Green Knight* which, as previously mentioned, is part of the exposition of the meaning of trawthe underpinning the pentangle on Gawain's shield.

- The *paz* of the whole manuscript, Line 1113 of *Purity*, raises the question of how the reader can apply the stories within the poem so that the mire of sin is removed and the righteous soul can become a pearl. (It forms one of the poet's characteristic linked pairs, being joined to Line 1111 which appropriately mentions God on his throne, symbolic of heaven and the City of God. It also mentions *mercy*.)

- The *paz* of *Patience* and *Pearl* together refers to reason not straying from justice.

- The *tau* of the *paz* of the second half of the manuscript (i.e. $\tau^3 \times 3062$) falls in the middle of a description of the Green Knight's armour with its gold and knots. Gold is mentioned on Line 190, knot on Line 194 and gold again on

Line 195 of *Sir Gawain and the Green Knight*. Curiously, Line 190 corresponds to a use of $^{21}/_{34}$ as τ and Line 195 corresponds to the use of $^{13}/_{21}$. It seems as if the author were hedging his bets with regard to the most accurate fraction.

Taking all these elements into consideration, it is clear there are many common elements which echo the sentiments of courtly love while at the same time being correctives to them. Leaving aside the paz of *Patience* and *Pearl* combined, to which we will return in the final chapter, and also leaving aside Jonah talking to the sailors in *Patience*, only the receiving line of the hunt and the animals at bay seem out of place. Yet perhaps even those fit neatly: vegetarianism in the Middle Ages was despised and linked in popular thought to Pythagoreanism[306] which frowned on eating meat because of the belief that human souls could be re-incarnated in animal bodies. So, although it's stretching a point to see the *tau* and *paz* positions which refer to hunting as definitively anti-Pythagorean in concept, it is not, given the overall tone of the *Pearl* poet's work, stretching particularly far.

That Chrétien's writings are the poet's main target is further evident in the lead-up to the Arming Sequence and the description of the pentangle. Ten knights are mentioned as attending the feast which Arthur gives on All-Saints Day. They are Ywain and Erec (both heroes of works by Chrétien—*Ywain, The Knight of the Lion* and *Erec et Enide* respectively), Dodinel de Sauvage and the Duke of Clarence, (who according to *Le Roman de Lancelot* were brothers), Lancelot himself, as well as Lionel, Lucan, and Bors, (all three Lancelot's cousins), Bedivere and Mador de la Porte.

Every knight mentioned is either connected to the tales of Lancelot or to the works of Chrétien. This feast scene occurs immediately before Gawain's departure and the names mentioned have clearly been chosen to prime the attentive listener or reader with thoughts of Lancelot just prior to the Arming Sequence. The final name in the list—Mador de la Porte—seems particularly designed for this purpose. In Arthurian romance, Mador is rarely mentioned. His fame rests solely on his fight with Lancelot over Queen Guinevere's honour.[307]

The purpose of directing the reader's thoughts towards Lancelot as the Arming Sequence begins is not clear until its end a hundred lines later. There, the poet finally reveals himself; devising as consummate an ambush as you could find anywhere in literature, a snare worthy of the devious Green Knight himself. Immediately after Line 666, in the middle of the thirtieth 'bob and wheel': the mention of a lance. In context, with the lead–up of Lancelot's kin and one of his most notable opponents, followed by the articulation of the pentangle's ancient lore—indicating it has a history quite apart from Pythagoras—the finale probably wasn't subtle in the fourteenth century. The poet was declaring a judgment call: in his view, neo–Platonic Pythagoreanism is a manifestation of the anti–Christ. Perhaps his presentation of Gawain as the *tulk* of *trawthe*—the man of truth, the interpreter of trawthe, faith and integrity—is to set up an antithesis for the 'man of lawlessness,' another name for the anti–Christ.

Nonetheless, this is all subtle enough for this particular interpretation to be open to question; however, in its favour is Chaucer's overt reference to Pythagoras with exactly the same line numbering in *The Book of the Duchess*. There, the sage of Samos is explicitly mentioned just after Line 666 in a poem that shows a golden ratio construction around the number 490 and is ambiguous with regard to the number 153. Is the *Pearl* cycle part of a reaction to Chaucer's increasing influence and slippery faith?[308]

Moreover, Line 666 of Sir *Gawain and the Green Knight* belongs to an interesting sequence. The addition of 1111+153+231+490+618 leads to Line 1391 of *Purity* where there is an important clue that these numbers are the key to the poet's architecture. At that point, the difficult word 'med' occurs. Andrew and Waldron are not sure what to make of it, beyond the fact it is a mathematical term; they suggest it might be a variant on 'mete', *proportion*.[309]

I suggest that it could also mean *equal, measure(d)* or *meetly fitting* and is a reference not only to the summation above but also to the position 17×153, which occurs only 2 lines earlier. Moreover

1618 lines beyond this lies Line 666 of *Sir Gawain and the Green Knight*—implying that the lance is not a mere lance but symbolic of deeper substance.[310]

All this suggests that those hunting scenes at golden sections really are conceived as anti–Pythagorean statements. This leaves only Jonah and the sailors as apparently showing no reference to either Delphi or neo–Platonism.[311]

Yet even this absence might make sense. It is clear from this single anomaly that *Patience* is not to be considered in its entirety but as a 'prologue and body': this is particularly evident when we note that the *tau* and *paz* positions of the body are 111 lines apart. If 1111 is a metaphor for the City of God or the Kingdom of Heaven, then we may suspect that 111 is the same.

This fits so very well with Jay Schleusener's suggestion that the fettling of poverty and patience reflects Augustine's commentary on the first and last Beatitudes as a meditation on the Kingdom of Heaven and 'a linear ascent to perfection and a cycle of progress which returns to its own beginning'[312] that I can only conclude the prologue has been specifically designed to be 60 lines long for the precise purpose of producing a 111–line difference between the two golden cuts. It thus builds into the poem (whose 531 lines already reflect a verbal circularity and a 'sum of perfection') the mathematical grammar of the Beatitudes. The whole poem is thus 'fettled' with links—literary, theological and arithmetical—between poverty and patience, truth and mercy, righteousness and peace.

This fettling is true for more than just *Patience*. In fact, *peace* is featured sufficiently often at golden sections—five times in all—that it begs the question of its significance. Five is a very suggestive number in the context of the golden ratio: it turns attention back to Gawain's shield with its pentangle of fivefold fives.

Yet, why *peace*? Is it a pun on Spanish 'paz'? Is it a pun on Hebrew 'paz'? Certainly a word like the *paz* that I've coined is likely to have been in use: it may originally have been *phares* or *pharez*.

618 lines from the beginning of *Sir Gawain and the Green Knight* marks the beginning of the digression about *trawthe* and the

pentagram. 618 lines backwards from the beginning of *Sir Gawain and the Green Knight* brings us to Line 1725 of *Purity*:

> *Without more ado, these are the words here written,*
> *with each character, as I find (it) as it pleases our Father:*
> *Mene, Techal, Phares*[313]

Here begins Daniel's exposition of the Writing on the Wall. As is usual with the poet, he puts an unexpected twist on what is well-known. In the Biblical account, it is mene which is repeated in the inscription *MENE, MENE, TEKEL, PERES—number, number, balance, divide.* The poet, however, doubles the word 'phares,' emphasising the importance to him of *division* over *counting numbers.*

Phares not only means *division* but is a pun on Persians. However, its homonym, pharez, means *fine gold.*[314] Pharez became modern Hebrew and Spanish *paz.*

202 lines back from the first mention of 'phares' in Line 1727 is the reference to the golden gods of Gaul. Yes, it's another multiple of 101 and the poet is pulling no punches with what he thinks about the denizens of courtly love: he names the gods associated with pride, evil and the dungheap.

2272 lines further back still, at the *paz* of *Pearl*, is that subtle reference to Daniel mentioned previously: the invocation of God as the 'Master of Spiritual Mysteries'. This is in fact 2475 lines from the opening of the *Mene–Techal–Phares* sequence which may not be quite perfectly 4 × 618, but is so close given the other mathematical constraints in play, that it seems extremely likely its proximity is not mere coincidence. Indeed, this positions the reference to the 'Master of Spiritual Mysteries' almost exactly 5 × 618 lines from the beginning of *Sir Gawain and the Green Knight* and 6 × 618 from the beginning of the description of Gawain's shield. This appearance of a multiple of 5, given the importance of five in the description of the shield itself, doubly emphasises the significance of the prophet Daniel in the poet's own thought about the golden ratio.

To highlight the importance of Line 1725 of *Purity* in relation to multiples of 618, it is noteworthy that 2 × 618 lines back is Line 489, abutting Line 490 which, not unnaturally, speaks of reconciliation and forgiveness.

This scene is therefore critical in its placement. The Writing on the Wall—mene, tekel, phares—is intimately linked in the poet's thought to the decimal digits of the golden section.

Mene is usually translated *number*[315] or *numbered*, however, it also has a sense of *appointed*, as in an appointment with destiny. This evokes the alternative suggestion of JJ Anderson for 'poynt' in *Patience.* The first line of *Patience* is, however, only 85 lines away and that numbering doesn't immediately suggest anything much. The last line of *Patience* on the other hand is 616 lines away and that does indeed suggest we ought to proceed cautiously and look for other evidence in case 616 is significant. 616 is close enough to 618 to get swamped in its halo—and that is the last thing that should happen because 616 was the *original* Number of the Beast in the Book of Revelation.[316]

Whether the poet knew of this earlier 'mark' is difficult to ascertain. 616 lines into *Sir Gawain and the Green Knight* is a description of the bright diamonds adorning Gawain's helmet. Is this a reminiscence of the 'radiant brow' of Taliesin, the Welsh bard who in a set of poems variously dated from the ninth to the twelfth century features as the final metamorphosis of Gwion Bach 'Fair Boy'? Gwion is a name so close to Gawain that the poet may have left us a clue that he knew exactly why the old Celtic hero waxed in strength in the mid-afternoon, or in some versions, increased in strength mightily at midday and midnight.[317]

Condren examines *Purity* in some detail, after noting that Daniel's interpretation of the writing on the wall (mane = number; thecel = weigh; phares = divide) corresponds to the famous passage in the Book of Wisdom where God is said 'to have ordered all things in measure, and number, and weight'.

He finds ten embedded constructions which use the golden ratio to develop the size of different sections relative to one another. He did not however examine the wording of the golden section of the whole poem, nor note that the writing on the wall begins 618 lines backwards from the start of *Sir Gawain and the Green Knight. Purity* has at this most crucial point a message to the mathematically–minded reader that the writing on the wall really pertains to the golden ratio.[318] The very same message underpins Lines 619–669 of *Sir Gawain and the Green Knight*. The latter expounds the nature of the symbol of *trawthe*, the former spells out the consequences of abandoning it. Those who act like Baltazar, defiling the golden vessels of the temple, will share his fate. Those who, like Gawain, are steadfast in honour will, even if they fall from the high standard which the pentangle symbolises, be forgiven.

In *Sir Gawain and the Green Knight*, there's a fettled relationship between 490 and 153 and the golden ratio. There appears, however, to be no such relationship in *Purity*. On the one hand, this seems a little surprising.

On the other hand, perhaps it is more appropriate to emphasise 153 in *Sir Gawain and the Green Knight*, not merely because of the resurrection theme detected by a number of scholars but because one of the traditional stories of Gawain bears a marked resemblance to the legend of Pope Gregory.

In some versions of the Gawain story, Gawain is the illegitimate son of a queen and a pageboy (later to become King Lot) who, at birth, is cast out in a coracle upon the sea. Like Gregory the Great in a similar story, he is found and raised by a fisherman, eventually making his way to Rome to come finally into an unexpectedly high destiny. *St. Erkenwald* also has a curious connection to the 'sinner made saint' pontiff: the story of the virtuous pagan judge in that poem has no known connection to the historical Erkenwald but it is very reminiscent of the story of Pope Gregory and the Emperor Trajan.

Moreover, as Aaron Wright has shown, 153 is hidden within the story of Hartmann von Aue's late twelfth century verse novella, *Gregorius*. Wright identifies its use as part of the exegetical tradition going back to Augustine which confers on it a status as 'sign and symbol of the body of the elect'.[319]

While I have my doubts about the influence of Augustine on the *Pearl* poet's mathematics, the use of 231 (with its aliquot factors of 153) in *St. Erkenwald* as well as the combination of 490 and 153 in

Sir Gawain and the Green Knight seem to signify that both poems have been inspired to some degree by this story of Gregory. Moreover, the use of the numbers in *Sir Gawain and the Green Knight* to reference the Golden Mass of the Five Wounds on the Cross reveals the poet's own thought about Gawain's perilous adventure: Gawain's true hope as a Christian knight ought to be in the resurrection of the body, not in a magical talisman that offers escape from death.

It seems almost unreasonable of the poet to have waited until 618 lines into *Sir Gawain and the Green Knight*—over two-thirds of the way through the manuscript—to pull out the first real hint that there was such an important golden ratio sub-structure built into the text, along with an arithmetic resurrection motif. Yet did he wait that long? Did he allude to it in the opening lines of the very first poem?

The first two lines of *Pearl* are:

Perle plesaunte, to prynces paye
To clanly clos in golde so clere

The second line clearly tells us that *Pearl* is set in gold. Is this a pun not just for the precious metal but for a mathematical and literary design, a design already more than a thousand years old, a design associated with the word 'gold' since at least the time of Chrétien's romances and possibly as early as the *Exeter Book*?

The clues to the underlying mathematical construction are relatively easy to follow. The *Pearl* poet has not been secretive—on the contrary, he tried his hardest to make sure all the signs were as obvious as possible.[320] Perpetual candles illumine his text.

He tells the reader at the very beginning of *Sir Gawain and the Green Knight* (Line 35) that his story is in the style of a lay told in the long-standing tradition of 'lel letteres loken'. Usually taken to indicate the revival of the alliterative form in the poem, it could just as easily refer to numerical literary technique which, after all, had a far older pedigree.

When did the poet write the *Pearl* cycle? It is comparatively easy to set the most probable date for the composition of *St. Erkenwald*— it was hurriedly written sometime in April 1386 to be ready for the first of the feast days announced by Bishop Braybroke at the end of that month. With this in mind, it seems extremely likely that in April 1386, the poet was already tackling the much more complex *Sir Gawain and the Green Knight* and possibly had even written the First Fitt. The mathematical structure of that Arthurian poem was at that point in time so familiar to the poet that it was a natural choice as the basis of the hastily flung–together *St. Erkenwald*. His thoughts were already mulling over how to re–work the legend of St. Gregory, while bringing in the number 153 and making clear the stark choice between Christian resurrection and the magic of the Pythagoreans who believed in re–incarnation.

He chose of course to make some sweeping changes while keeping to those same themes; however, the alteration from a triple triangle planned for *Sir Gawain and the Green Knight* to the triply triangular number of *St. Erkenwald* is a touch that didn't require much thought. It's an idea which had already been around for some decades as indicated by *Sir Orfeo*.

How much he was influenced by Geoffrey Chaucer is difficult to tell[321] but, if he knew Chaucer's work, it is certain that he would have been professionally aware of the mathematical structure underlying the *Book of the Duchess* and *Parliament of Fowls*, just as in a later age Milton seems to have detected neo–Pythagoreanism in Shakespeare.[322]

If the *Pearl* poet did not know either Chaucer or his work, then some very strange coincidences have occurred. Recall that Stanza 17 of *Parliament of Fowls* is, as Condren has suggested, structurally dispensable.

In a parallel fashion, there are 17 lines from *Pearl* which are structurally dispensable. They occur in Section XV in the very part of the poem which has always attracted attention for its irregular nature. Although it is unnecessary to propose the deletion of 12 lines, as editors often have in the past in their desire to have precisely 100 stanzas, the anomaly at this point (6 stanzas instead of 5) suggests that we take a closer look at the deep structure. Lines 868 to 884 can be removed from the text without damage to the sense. The rhyme suffers slightly but the concatenation does not. So since these 17 lines can be removed and the text remain readable, it should be noted that there are additional clues pertaining to 17: the line immediately before these is Line 867 (or 51 × 17 = 3 × 17 × 17) and 17 itself is τ^3 of 72 which is the number of lines in this already anomalous Section XV.

It seems that Lines 868–884 are probably hiding something of major significance.

Anyone obsessed with 1200 lines or 100 stanzas as the true presiding numbers for *Pearl* will certainly not agree that 17 lines can be shunted neatly aside. However, looking at the overall manuscript, a loss of 17 lines would mean that that total is 6069 lines, instead of 6086. This is still divisible by 17, of course, but now it's also divisible by 17^2 and 7 and 3. If the poet encoded his name in the manuscript, and I do believe that this is still a very big *if*, it would not surprise me if it were found in this particular section.

Other anti–Pythagorean thoughts may have pre–occupied the poet's mind as he devised *St. Erkenwald*: it is possible that the intricately–layered allusions of the Arming Sequence of *Sir Gawain and the Green Knight* were constructed entirely separately to the poem proper, long before the rest of it was created. If that is the case, then the subtle allusions to Delphi built into that part of the poem may have been so pre–eminent in the poet's thought that they segued across into *St. Erkenwald* as well.

Indeed the Arming Sequence of *Sir Gawain and the Green Knight* is dense with allusion and architectural splendour. Its flying buttresses spread in several directions. We have already noted that the reference to the Five Wounds on the Cross falls on Line 643 instead of 638, apparently to connect a golden section to the end of the first Fitt through the resurrection number 153. There are other constraints that pull perfect mathematical metaphors ever so slightly out of alignment.

490 lines beyond Line 618, Sir Bertilak suggests that the 'exchange of winnings' game be agreed on with *trawthe*. This is not, as Andrew and Waldron suggest, merely to offer an honourable answer, it is on the contrary the most binding of oaths—the swearing by God. This is two lines from Line 1111, the kingdom–of–heaven symbol.

Indeed, an effort to link the metaphors of the City of God through forgiveness to trawthe may have determined the first mention of the pentangle, which is on Line 620 when it would have been far more mathematically appropriate to have it one or two lines earlier.

Before leaving the controversial topic of the *Pearl* poet's wider oeuvre, I must not fail to mention *Death and Liffe*. The author of this undated poem has been so greatly influenced by the *Pearl* poet's mathematical grammar that it must be suspected the original of *Death and Liffe* was actually written by the *Pearl* poet himself. Not only are there a number of expressions for God found nowhere other than the *Pearl* manuscript, but the poem is 'circular'. Moreover, the 459 lines of the poem are divided into 2 fitts which are almost perfectly split according to a golden section. The one line difference necessary for a perfect split is clearly constrained by the fact that the poem, at 459 lines rather than 460, is divisible by seventeen.[323]

The difficulty with dating *Death and Liffe* has been brought about by another poem of much inferior quality but remarkably similar compositional style: *The Scotish Feild.* This, a stirring account of the Battle of Flodden in 1513, was written by a poet who described himself 'as "a gentleman, by Jesu" who had his "bidding place" "at Bagily" (*i.e.* at Baggily Hall, Cheshire).' He was 'an ardent adherent of the Stanleys and wrote for the specific purpose of celebrating their glorious exploits at Bosworth Field and at Flodden. The poem seems to have been written shortly after Flodden, and, perhaps, rewritten or revised later. That the author of this poem, spirited chronicle though it be, was capable of the excellences of *Death and Liffe*, is hard to believe; the resemblances between the poems seem entirely superficial and due to the fact that they had a common model'.[324]

Although Sir John Stanley[325] is sometimes considered as the author of *Sir Gawain and the Green Knight*, it should be noted contrariwise in favour of John or Hugh de Massey that Baggily Hall was merely three miles in one direction from Stockport and four miles in another from Ashton–on–Mersey. Moreover the Baggileys[326] appear to be a branch of the Massey family,[327] since John Stanley's mother was in fact Alice Massey. It is therefore more likely that the author of *The*

Scotish Feild is a descendent of the Leighs who inherited the property when, around 1353, it was made over to Sir John Leigh of Booths, a widower who married the heiress, Isabella de Baggily. In this scenario, *The Scotish Feild* author simply copied the distinctive style of *Death and Liffe*, a manuscript which had been originally composed over a century earlier and which was preserved in the family as a unique copy.

It seems likely to me that there was never more than a single manuscript of *Death and Liffe* or the *Pearl* cycle. The survival of these poems is therefore in many ways miraculous. That the works were solitary copies may well be accounted for simply by bad timing, given the nature of their themes. They are all theological in essence and, more importantly, written in the vernacular—a combination which spelt a death–knell to wide dissemination, if the timing of their completion was the first decade of the fifteenth century. In 1409, the *Constitutions* of Thomas Arundel, Archbishop of Canterbury, forbade vernacular expression of theology.[328] Given the deep underlying complexity of the *Pearl* manuscript and the likelihood that it took decades to put the last touches on the entire poem cycle, it is conceivable that it was finished not long before Arundel's interdiction came into place.

However, the *Pearl* poet is a man of such intelligence that I do not think he was ever limited solely to the dialect of the English Midlands. His mathematical style is so distinctive that it would not be surprising one day to find it turn up in Latin in a poem which has a theological theme: to ask him to write vernacular poetry without religious content was, I suspect, about the same as asking a bee to produce milk instead of honey. If it was a choice between the vernacular and theology, I doubt there was a moment's hesitation on his part. One can only hope that, should such a poem ever come to light, it would have the poet's name clearly and unequivocally attached.

In conclusion, the only major difference between the complex and ornamental mathematics of *St. Erkenwald* and that of *Sir Gawain and the Green Knight* is a vast simplification of the musical overlay. *St. Erkenwald* is blunt while *Sir Gawain and the Green Knight* is subtle; they share an apparently common inspiration which welds disparate themes: an admiration for Pope Gregory (or at least his legend) and a desire to see the symbols of Delphi reduced to rubble. It seems to me that the reason for these common themes was that the author cannibalised his own ideas to make it possible for a sudden, unexpected commission to be completed on time.

Perhaps the language differences noted by Larry Benson can also be accounted for by this very point—lack of time. As any halfway decent writer of fiction will tell you, the difference between one manuscript draft and the next is often so monumental as to make it seem like entirely different people wrote them.

digression 5

The lord of Hautdesert, Bertilak the Green Knight, has a curious turn of phrase each time he proposes a bit of festive sport. Although he purports to deliver his challenges in the spirit of a party game, he uses the language of the law to bind Gawain into a legal agreement. Not only does he repeatedly use the term 'covenant' in relation to his wagers, he makes sure Gawain repeats the terms of the pact, sometimes more than once, in order to ensure it is clearly understood. At Hautdesert, he even calls for a drink to seal the 'covenant'.

The use of this word is so frequent that, for some commentators, it calls to mind the Old and New Covenants: that is, the Old and New Testaments of Christian theology. There is an emphasis of the letter of the law and strict adherence to the terms of the covenant in the first instance and grace in the second. This reflects a move in Bertilak's attitude: from a knight who requires formal justice to one who tempers strict justice with compassion and mercy.[329]

According to this interpretation Bertilak embodies at least two of the virtues of the Four Daughters of God at different times. There is much to recommend this perspective. However, I believe it overlooks a significant point.

In Hebrew, a covenant is *cut*, rather than *made*. A child of the covenant is brought into that relationship by the sign of a cut in his flesh: circumcision. The 'cut' that Gawain delivers to the Green Knight and its reciprocation both occur on January 1, the Feast of the Circumcision of Jesus. Thus Bertilak's constant references to a covenant may belong to a wider background—one that acknowledges that the 'cut' in Gawain's neck has a deep Hebraic concept behind it.

A formal blood covenant in ancient Israel had a number of ritual aspects, including the passing of the covenanting parties between the partitioned body of a slain animal.

The structure of the Fitt III of the poem continually places Gawain's story between episodes of a daily hunt, positioning it neatly between descriptions of animals that are killed and cut into pieces. This design recalls the most ancient symbolism of a covenant.

Linda Bates draws attention to this when she points out how in 'a simple image of the hunt, the context of the preparation of the carcass is resolutely concerned with covenant. As the men "hewe hit in two" the reader is reminded both of the physical gift that Bercilak will make to Gawain (his "sacrifice" to his guest) and of the bargain they have made.'[330]

Indeed, the exchange–of-winnings game may reflect the exchange of gifts which was an aspect of the blood covenant rite. If it was the poet's intention to point towards this ancient Jewish code, the relationship between Gawain and Bertilak is entirely different to any presently posited. A blood covenant brings the participants into a relationship deeper than that of brothers: they are 'one' with each other, taken into each other's family, sworn to defend each other to the death.

Like David, the armour–bearer of Saul and therefore his covenant defender, Bertilak cannot raise his hand to kill his covenant–pledged brother–in–arms.[331]

It may be that this suggestion is taking circumstantial evidence too far.

Perhaps the covenant is linked, as previously noted, to Aeneas. Or perhaps it's meant to evoke both.

Chapter 7

The Kiss of Heaven & Earth: the number 1743

Kissing with golden face the meadows green,
Gilding pale streams with heavenly alchemy...

William Shakespeare

Before we re-examine the most momentous of the poet's battles against Chrétien, it seems a good time to pause a moment and examine the era in which he lived. Permit me to indulge for a page or two in one of my other passions besides mathematics: meteorites.

The fourteenth century was a time of great social and political upheaval. Wars, rumours of wars, plague, signs in the sky: all the right elements for imminent apocalypse were there. It may be difficult to pinpoint one particular matter amongst so many, but nevertheless it's worth asking what fear would have been uppermost in the minds of men and women at the time. The Great Plague had ravaged through Europe in the middle of the century and, consequent upon it, much social unrest because of the shortage of labour. No doubts in the minds of many people, the plague had been a turning point in their own fortunes as well as that of king and country.

Although this is greatly simplifying the problems of that turbulent era, it can be said without qualification that there is little doubt the greatest general fear of the time was the possibility that the plague might recur. It therefore goes without saying that the *perceived* cause of the plague was a matter of extreme concern. It is not really relevant what the real cause was, only what the people *thought* it was.

At the very beginning of the century, in 1301, Halley's Comet appeared. Two years later, Giotto painted the nativity of Jesus, depicting the Star of Bethlehem as a comet. Halley's Comet returned in 1378, the year after the death of Edward III. It was not nearly as bright as in 1301, passing at perihelion within $10°$ of the north celestial pole. Nonetheless such an omen would not have gone unnoticed since, as Napier and Klube have shown, there was an immense upsurge in the late fourteenth century in long-period comet-spotting activity. The *Pearl* poet indeed has his eye on the north when he has Lucifer speak on Line 211 of *Purity*, declaring his intent to set up his throne on the 'tramountayne'—*the pole star*—so called in Italy and Provence because it was visible beyond the Alps. While this may simply be an allusion to the traditional direction with which the devil was associated,[332] it may also have an association with the fall of a comet.

Mike Baillie, whose work on tree-ring dating has led him to look at comets as a possible cause of cataclysmic climate change in historic times, attributes the sudden rise in comet-spotting in the fourteenth century to the widespread belief that plague was caused by 'pestiferous vapours' associated with comets. Indeed the Black Plague of 1350, the cause of the most widespread social disruption of the century, was attributed at the time, not to fleas and rats as is generally theorised today, but to the presence of a noxious black[333] comet which closely grazed the earth.

The name 'comet' means *hairy star*, from Latin 'coma,' *hair*.

How often are there mysterious references to hair in medieval romance? Chrétien of course simply couldn't leave the idea alone. Lancelot in *Le Chevalier de la Charette* swooned constantly at the sight of Guinevere's golden hair on a comb while in *Cligés*, Sordamour's golden hair was sewn into a shirt.

Furthermore according to Roger Loomis, Gawain's name, it may be recalled, comes from 'Gwallt Avwyn' meaning *hair like reins* or 'Gwallt Advwyn' meaning *fair-haired*. Is this just coincidence or is Gawain not a solar hero, but a comet? Is there an allusion to Taliesin's 'radiant brow' in Line 616 of *Sir Gawain and the Green Knight* and is this an indication of the poet knew both of these Arthurian heroes to have their origin as comets?[334]

Mike Baillie, at least, suggests that the legends of Arthur and the wasteland should be seriously considered as a dim memory of the catastrophic climate collapse which is recorded in tree rings around the globe and which occurred in the years 536–540. He also suggests that the legends of Apollo, as well as many Celtic gods, make much more sense when seen as pertaining to a comet, rather than the sun.[335]

Baillie points out that history has been too often dismissed as legend, simply because it seems too incredible from our modern rationalist viewpoint. On realising that tree rings he was examining showed signs of a prolonged devastating winter in the mid–sixth century at the very time traditionally associated with Arthur and Camelot, he turned to historians, only to be told repeatedly that there were indeed records about an unnaturally long winter, but they were untrustworthy and had long been dismissed as fabulous exaggerations.

The winter landscape Gawain passes through on his way north has long been compared to the 'northern European Waste Land, the Utgard of Norse mythology'[336] which is then compared to a bitter winter in Wales. The comparison perhaps is more appropriate to the enchanted Fimbulwinter brought on by the giants, who as jotuns are clearly related to the 'half–ettin' Green Knight. As Baillie points out, perhaps the Fimbulwinter is not myth at all but an account of the devastating after–effects of an asteroid or comet strike.

Although Baillie does not mention the following in his examination of evidence for climate collapse over the centuries, I think it summarises the problem perfectly. The last sentence in the following translation of the *Anglo–Saxon Chronicles* for A.D. 793 is perhaps the most famous in that entire record. The sentence before it is almost always excised, lest it disturb the sober reliable tenor of the whole.

In this year terrible signs were seen in the skies over

*Northumberland and people were horribly afraid. There was
incredible lightning, and fierce winds, and fierce dragons could
be seen flying through the air. And right after this there was
a great famine, and not long thereafter, on the eighth day of
June, alas! plundering heathens destroyed God's church at
Lindisfarne, robbing and murdering.*

(tr.) Burton Raffel

Incredible lightning, fierce winds, famine. It all sounds realistic
except for those dragons. Yet this passage has all the hallmarks of a
large meteor strike followed by an impact winter and accompanying
harvest failure—a failure which was far from local given the sudden
appearance of the Vikings.

As I could use the term 'mushroom cloud' without ever meaning
sky-borne fungus, I don't doubt 'dragon' could be a colloquialism
used by the chronicler for the multi-coloured worm-like ionisation
trail of a large meteor.

All this means, as Baillie has pointed out, that we need to be
perilously careful in examining our own assumptions. Given that
Pythagorean mysticism has always been linked, not just to a religion
of numbers, but to worship of Apollo, is it possible that Gawain is
a counterpoint to celestial references in the works of Chrétien?
Moreover, do those works conceal neo-Pythagorean references to
Apollo as both a god and a comet?

Are there, possibly, allusions to comets in *St. Erkenwald* as
well? Two of the sub-divisions of that poem are, as previously
mentioned, 320 lines and 112 lines, which conceivably correspond
in the context of Harold Godwinson to the years 1066 and 1178: the
first corresponding to arguably the most famous re-appearance of
Halley's Comet and the second corresponding to an event raised out
of obscurity by Isaac Newton—the report of the sighting of a lunar
impact by five monks of Canterbury.[337]

Whether the poems do contain these particular allusions or not,
one thing is certain—both poets have laboured hard at incorporating

into their works astronomical references to the sun and moon, as well as to the measurement of heaven and earth.

Joan Helm's dissection of the mathematics of *Erec et Enide* provided a wealth of information which demonstrated *Pearl* poet's response to, and loathing of, Chrétien's neo–Platonism.

For the *Pearl* poet, the meeting place of heaven and earth was not to be found in Chrétien's esoteric measures with their disturbing neo–Platonic overtones and belief in re–incarnation, nor even in the 'thin places' of the Celtic twilight where the otherworld presses close to everyday reality. His own works may evoke that dream–like borderland of his Welsh neighbours as well as the literary dreams so common to medieval poetry like that of his countryman, Chaucer, but for him the mystical union of heaven and earth is in the kiss of peace and justice, and in the caress of mercy and truth.

Peace, justice, mercy, truth: the Four Daughters of God have, as we have seen, carefully positioned mathematically.

- *Trawthe* and peace have been repeatedly stressed at golden sections.
- Justice has been emphasised in *Purity* and mercy in *Patience*, the two poems that meet in the middle of the manuscript.
- The pentangle of *Sir Gawain and the Green Knight* is symbolic of truth, faithfulness, integrity, justice, righteousness. It's almost a meeting place of the Four Daughters of God in its own right.
- God speaks to Jonah 490 lines into *Patience*, discoursing on mercy and forgiveness—this is mathematically elegant, not simply because 490 is linked to forgiveness again, but because this position marks $^{153}/_{265}$ of the way through the manuscript, highlighting the $^1/_{\sqrt{3}}$ or the 'measure of the fish' so important to the early Christian believers in the Resurrection.
- For his mathematical token of mercy and truth,

the poet combines 490 and the golden ratio.
For the token of the kiss of peace and justice
he uses 531 and zero.

Let's see how he confirms for us that this is just what he has done.

The fettled Beatitudes of *Patience* are based on the Sermon on the Mount with its instruction on the coming of the Kingdom of Heaven. This Kingdom is portrayed, as previously seen, in the New Jerusalem descending from heaven at 231 lines from the end of *Pearl*—231 being symbolic both of resurrection (since its aliquot factors add to 153) and of trawthe (since it is an arithmetic triple triangle.) However, 231 lines from the end of *Pearl* is Line 981, which happens to be adjacent to a multiple of a token of forgiveness, Line 980.

But the Kingdom is more deeply fettled.

Remember that *Patience* can fold up to zero, making a pleat in the fabric of the manuscript. Remember too that the pleat can extend 265.5 lines in either of two directions—that is, it can touch as far 265.5 lines forward into *Sir Gawain and the Green Knight* or 265.5 lines back into *Purity*. A symbol of peace features in *Sir Gawain and the Green Knight* while the coming of justice appears in *Purity*. The holly bob[338] of the Green Knight—the evergreen symbol of peace—is mentioned in the former and the horror of the king of Babylon as he looks at a mysterious writing penned by a disembodied hand is detailed in the latter.

It's almost impossible to keep justice and mercy in equal psychological balance, so the fact that the poet harmonises the extremes he has portrayed in *Purity* and *Patience* is a remarkable testament to his faith. Judgment and blessing, righteousness and clemency come together in the astonishing ending of *Sir Gawain and the Green Knight*. The final challenge faced by Gawain at the Green Chapel is an exemplar of both justice and mercy delivered simultaneously. In the scene where Gawain's head is on the chopping block and the Green Knight takes three swings of the axe in order, eventually, to nick Gawain's neck ever so slightly, the demands of both justice and mercy are met simultaneously. The blow that draws blood requites the call of justice; that the same blow does not kill requites the demand of mercy.

The *Pearl* poet has produced a theological outcome of remarkable skill and unquestionable insight—the fact that this balance is hardly achievable in the human mind with its resolute preference for one or other of God's attributes but not both together is a testimony to the depth of his Christian faith.

The poet has resolved the dilemma and produced a psychologically satisfactory and incidentally spiritually edifying ending to his cycle.

Moreover, justice and mercy flow out back to his opponents: Chrétien de Troyes, Jean de Meun, Guillaume de Mauchat, perhaps Geoffrey Chaucer. He is forthright in everything he detests about their work, but courtesy is his hallmark in his treatment of it. He deserves the deepest admiration for his achievement. Chrétien veiled what he wrote with the most cryptic of clues—but the *Pearl* poet went out of his way to draw the educated reader's attention to his numeric devices. What is the difference between numerology and numerical metaphor? One is an act of magic, the other an act of investing meaning. In terms of the difference between the two poets, one is a spell and the other is a prayer.

Countryfolk of the nineteenth century were well aware that it was almost impossible to tell the difference between the highest forms of black magic and prayers. It would not surprise me in the least if it is eventually discovered that Chrétien's romances have a lot in common with incantations and that magic is in their very bones.

The whole of *Sir Gawain and the Green Knight* is ultimately directed towards the opposition of faith and magic: given the poet's pre-occupation with the works of Chrétien, it seems likely there is much more of the dark art in his work than generally recognised.

Because like Chrétien, he seems to have been acquainted with Hebrew learning.[339]

One of the subtle touches which suggests this is found in *Purity*. In what seems to be a creative departure from the book of Genesis, the poet makes a link between Lot's wife turning into a pillar of salt and her previous 'sin against salt' during the visit of the angels to Sodom. This idea of a 'sin against salt' is part of rabbinical commentary.[340]

Another subtle touch is found in the circularity of *Sir Gawain and the Green Knight*. More like the concentric rings of Hebrew poetry, the structure shows a movement through four elements at both the beginning and the end: Troy, Brutus, Arthur and then the Round Table.[341]

What convinces me that the poet knew Hebrew is not so much these tiny touches as an even more subtle matter related to the overall theme. Embedded in his mathematics is a remnant of a Judeo–Christian belief that seems to have long disappeared. I have searched in vain to discover in any commentary of any era which contains a reference to this idea, so the poet seems to be unique in his use of it.

This remnant is found in the golden sections. The mathematical tokens joining so much of the manuscript are so obvious that the fact that two of the most important golden sections had no clear architectural connection was, at first, completely baffling.

The *paz* of the last two poems combined falls on Line 638, in the middle of the Arming Sequence and the description of Gawain's shield. It was always a pity that this was not 20 lines less and so that it fell on Line 618. But if the whole of the digression about the pentangle is of zero dimension then, in a loose sense, the *paz* does indeed fall on Line 618.

However, within the description of the pentangle, the virtues of justice, truth, integrity, faithfulness, loyalty, courtesy, pity all meet— as we look at them, we might consider them as euphemisms for the Four Daughters of God who met together at the previous golden section: the *tau* of *Patience.*

The kisses Gawain exchanges with Sir Bertilak—1, then 2, then 3—build like triangular numbers; these kisses reflect the summative structure of *Patience* discovered by Ed Condren.

The kiss of *Patience*, the armour of Gawain (with its reference through the virtues to the same kiss), followed by the kisses of Gawain and the lady and then Gawain and the host build up until there's no subtlety left: it's all too intensive. Before very long, it felt like a clue and not a simple re–inforcement. Is the *tau* of *Patience* with its reference to a kiss hinting at more in the next nearest golden section which already discusses armour in extravagant and expansive detail?

So: *when is a kiss like armour?*

It was a riddle too difficult to answer until I realised it was the wrong question.[342] '*Like*' does not really come into it. It's not a simile. When does being kissed *equate* to being armoured?

No wonder the poet emphasised Gawain as a tulk: we are about to boldly go, as far as I can tell, where no interpreter has gone before.

In his epistle to the Ephesians, the apostle Paul famously described the armour of a Roman soldier. In 101 Greek words,[343] he parallels it with the armour of God—those elements a Christian should wear to be protected against the wiles of the enemy: the belt of truth, the breastplate of righteousness, the shield of faith, the shoes of the gospel of peace, the helmet of salvation, the sword of the Spirit and songs of praise.

Those first four elements are fundamentally the Four Daughters of God.

Is this a coincidence?

I don't think so.

Nor do I think it a coincidence that Gawain's fault—inevitably against *trawthe*—consists in accepting not a ring or a sleeve but an allegedly magic girdle, a perfectly diabolic counterpoint to the belt of truth.

If there is any exegesis where the Armour of God (and thus the fivefold fives of Gawain's shield) correspond to the virtues of the Four Daughters of God, I have not found it.[344] Burrow mentions that 'allegorical elaborations of the Pauline armour of God were common in contemporary religious writings'[345] while making the point that modern readers find the connection between the pentangle and Gawain's moral virtues somewhat devious but that people in the late Middle Ages were accustomed to the parallel. However, he makes no connection to the Four Daughters.

In view of this complete failure to track down any contemporary source, I felt compelled to sideline my mathematical prejudices and check the gematria of Ephesians 6:12–18. Before doing so, I hypothesised that, if there were indeed a connection, the gematria should add up to 77777. I based this theory on two considerations:

(1) the armour of God has 7 elements

(2) the kiss of heaven and earth—at least by medieval times— was a run of ones like 111 or 1111.

I recognised that back–dating the mathematical motif of the 'mystical union' was a perilous notion. So the fact I was not correct in my guess wasn't much of a surprise. However, since the total could have been any number at all, it was surprising that it wasn't much wrong: the gematria of all the Greek letters in the sequence[346] turned out to be 77791.

The difference of 14 is miniscule in percentage terms—less than 0.02%—equivalent to 14 cents in over $777. If this were the predicted result of either a scientific experiment or a hypothesis test within statistical analysis, would be dismissed as completely trivial. But this is not a scientific experiment and, however small the error may be in terms of modern Greek rationalistic thought, it is not insignificant in Jewish thought. We are dealing in fact with the alphanumerics of a Greek text written by a Hebrew apologist of the first century: Paul of Tarsus hasn't given anyone a difference of 14 to play with; at most I get a difference of 1.[347]

And I intend to make use of that difference as I note that 77792 is divisible by 17.

Not only that, it is also divisible by 22.

It is the closest number to 77777 which is divisible by both 17 and 22.[348]

I don't understand what this meant to Paul and I hesitate to suggest any sort of correspondence between his motives and the *Pearl* poet's. However, I do raise the question: *Is this where the poet got his mathematics? Is this the source of his inspiration?*

I contend that, at the end of the day, it has to have been. And I conclude this, despite the fact that since 77792 is also divisible by 52 and 4, that a very similar kind of mathematical design occurs in John 1:1 where there are 17 words of 52 letters with two embedded groups adding to 777 and one adding to 707 within the gematria.

As far as I am concerned, when inspiration is, in effect, only negative (that is, the poet loathes courtly love so much he feels compelled to take up a pen to write the opposing view), polemic and diatribe occur.

Admittedly there could be some of that. At the end of *Sir Gawain and the Green Knight*, Gawain lets loose with a misogynistic outburst, displaying a hatred of women paralleling and perhaps even over-matching his previous courtesy towards them. These views are often taken as the poet's own sentiments. However, whether they truly are is debatable. If the extreme sentiments of *Purity* are the result of taking a particular rhetorical stance for the sake of argument, then surely it is likely this is the same.

Our poet is a tulk of rare sensibility and discernment. Perhaps he was unveiling a concealed truth long before modern psychology reached its own conclusion: behind courtly love and its elevation of women to a pedestal, there lies not respect but misogyny, not esteem but antipathy. Perhaps the poet puts this very idea into Gawain's mouth to reveal that courtly love is a mask, hiding disdain for women behind deference to them.

The odd thing about the poet's work is that, even while tending towards extremes for the sake of the debate of the Four Daughters, it still shows a marked generosity of spirit, even towards his opponents. This is why I contend that the poet was looking towards Paul's epistle. It's extremely hard not to produce polemic when you have only a negative role model and not a positive one. It's much easier to be generous when you feel confident of your backing. You don't need to bring out the attack dogs or adopt a defensive position, you can instead be the epitome of graciousness.

However, this notion that the poet was looking towards the letter to the Ephesians does bring certain implications with it:

- He has to have access to a Greek text and an accurate one at that.
- He has to have been able to read that Greek text sufficiently well to have been able to know where the description of the Armour of God in Ephesians begins and ends.

- He has to have known the alphanumeric value of each Greek letter.
- And most importantly, he has to have known that *nashaq*, the Hebrew verb for 'to put on armour', is identical for 'to kiss'. They are not just spelt the same way—they come from the same root and have the same meaning.

This last is quite stunning. Because I can't find any commentary—modern or ancient—that mentions it. Hope Traver comments on it but only in a footnote. If I had not suspected that the poet knew Hebrew, I would not have checked the Hebrew word for 'armour'. I would not then have noticed that it also meant 'kiss' and realised that Paul was referring to Psalm 85:10 in his description of the Armour of God.

This could have been common knowledge once upon a time in the fourteenth century, common knowledge that the poet put to good use.

But if it was, it has disappeared entirely in the last six hundred years.

I'm inclined to think, however, that the relationship of armouring and kissing[349] has not ever been common knowledge. Our poet is a tulk who can take a remez and weave it into his own work.

And perhaps, herein, is the reason why the poems of the *Pearl* manuscript were never widely disseminated.

Cheshire's fortunes waned when Richard II died in 1400 and Henry IV took the throne. Shortly thereafter the rebellion of Owain Glendower began—the Welsh marches rising against the English king. The hero of the *Pearl* poet's last poem has a name that is basically the same as Owain.

It would not have been a good political move to release a poem cycle into English court circles using a language favoured by the previous king and featuring a hero who has the same name as a rebel. By the time Glendower's rebellion petered out several years later, the language of the English midlands was even more out of favour. More importantly, however, there must have been rumours of unrest in ecclesiastical circles,[350] culminating in 1409, with the interdiction of Thomas Arundel on the writing of theological works in the vernacular and restricting them totally to Latin.

This, to me, spells out why the *Pearl* manuscript remained hidden for so long and was apparently never known in its own time.

It is theological at its heart and in its bones. It is the debate of the Four Daughters of God writ large. Each is featured in her own poem, various pairs team up to argue their points, all of them congregate at golden sections, and the entire cycle culminates in a pas de deux of mercy and justice when Gawain survives the ordeal at the Green Chapel.

The four poems throb with so much more than a general religious sentiment; from beginning to end, wherever their skin is scratched, what flows out from the verse and the mathematics is a weighty exegesis of Scripture.

We have already seen that the *Pearl* poet had a churchman's respect for the ban on tournaments. There is no doubt he would have had a churchman's respect for Arundel's interdiction.

The one and only copy of the *Pearl* manuscript—the intense and concentrated labour, no doubt, of decades—went into obscurity in the library at Baggily Hall or perhaps some abbey in Cheshire or Lancashire. And there it stayed until around the time of the dissolution of the monasteries under Henry VIII.

So who, after all this, was the *Pearl* poet?

Is his name encoded in the text?

Having considered all the evidence gathered so far, I'm inclined to doubt it. If he did embed it, I think it's most likely to be found somewhere in Lines 868 to 884 of *Pearl*. However I still think it's more likely he refrained, not just from mentioning his name, but from incorporating it into the text in any numerological fashion. If this is correct, then his work was part of that tradition of art offered in praise of God, not for the applause of the world.[351]

If in addition he was one of the de Massy family, it is ironic on several levels that in 1438 Piers de Massy was killed in a tournament, run through the head with a lance.

He was, however, regardless of his family background a man of towering intellect, a 'renaissance' man long before that time. His interests ranged across music, art, mathematics, theology, astronomy, hunting, poetry, architecture, history. His sense of fun is equalled by his intensity of purpose and passion for purity. His graciousness is unlike that shown by Chaucer: he is forgiving, where Chaucer is tolerant. Chaucer specialises in comic characters, drawn true to life; the *Pearl* poet in delightful wit, forged in a crucible of zeal. His irony is nevertheless flippant at times, occasionally sounding completely at home within the cultivated insouciance of the late twentieth and early twenty–first century. 'What?' Gawain asks the Green Knight in the most tense scene of the whole poem. 'You haven't frightened yourself with your own threats?'

Despite never taking his eyes from neo–Platonism, the poet always made sure his obsession did not come between him and a good story, boldly told. His genius seems to have been equalled only by his compassion and generosity of spirit. He was a man of immense commitment to an ideal who could, at one and the same time, take life with deadly seriousness and without any seriousness at all.[352]

Man is the measure of all things, said the Greek philosopher Protagoras.[353]

Not so, the *Pearl* poet clearly declaimed, over and over again. The measure of all things begins with the kiss of heaven and earth and with the coming of righteousness and peace. Into fourteenth century fears about earth–grazing comets bringing disaster and plague, he delivers a message focussed not on God punishing the world, but of caressing it with peace and mercy. He makes sure his mathematical metaphors flow constantly back and forth through *trawthe*, a word which may still echo faintly in modern mathematics.

Where did Martin Ohm get *tau* for the golden section? It doesn't just mean *cut*, it's also the Greek letter for both *tree* and *cross*. Just coincidence? Or does *tau* actually go back to all those medieval poems in which trawthe, treuthe, trwe and troth appear at golden sections?

Whether it does or not, *tau* seems a fitting successor to the most fought–over mathematical metaphor of all time.

In conclusion, I have to say that I've only just scratched the surface of what I believe is to be found in the *Pearl* manuscript. If I have any word of advice for anyone inspired to look, as I have, at the fusion of word and number which is ubiquitous in medieval poetry, it is this: simply remember the Grail Question: *What is the meaning of this?*

Mathematics in poetry from the classical to the medieval age isn't 'just there'—it is purposeful and meaningful, although sometimes enigmatic.

When I first learned algebra, I was taught it as a symbolic language. For many years, that was the way I introduced students to it—as did my colleagues. Then came the time when such pedagogy fell into disrepute. Experts not only expressed their contempt for the approach, they also despised the 'bad' mathematical thinking behind it and advocated the use of number patterns to build a solid foundation in pre–algebra.

As a result, I belong to that lost generation of teachers who have both a strong facility with numerical patterns and also with symbolic language. I tend to instinctively link the two. So, for me, there is no real effort needed to make the final extrapolation when it comes to patterns in numerical design: arithmetic metaphor.

If, in algebra, it's possible for symbols to stand for numbers, why can't the reverse also be true? Is it possible that to the medieval mind numbers could be symbols, pointing not just to mathematical concepts but to theological ones?

It's not really a huge stretch of imagination. Just one of philosophy.

The twenty–first century is not the fourteenth. We recognise this but give lip service to it in practice. We analyse the poetry of the fourteen century, assuming a shared worldview when it comes to mathematics. Nothing was more evident to me when I saw how many commentators considered that a line had been lost from Chaucer's *Parliament of Fowls* and that it was obviously intended to be 100 stanzas of 7 lines each in rhyme royal. But what if the missing line was a clue to the design? What if 699 lines was intentional—comparatively subtle when compared to poems like *Divine Love* or *Summer Sunday* but nevertheless a deliberate giveaway to the poetic game?

As it happened, at the same time as the change from symbolic language to number patterns occurred in the pedagogical approach to algebra, the overall curriculum went through one of its long–term cycles. Invariably the single most popular recurring concept in these periodic crests and troughs is the idea of the 'integrated curriculum': a proposed merging of subjects in rich, meaningful ways. In schools, the push against the fragmentation of subjects into ever smaller specialities—a process which culminates in tertiary study of ever narrower areas of expertise—never lasts very long. As it becomes evident it's too hard an ask from teachers who themselves are a product of the process of specialisation, it's quietly dropped after a year or so.

I started my examination of medieval poetry during one of these periodic pushes for a high–level integrated curriculum. I happened to be actively looking for examples of the golden ratio in literature. The golden ratio is hugely advantageous as an overall theme for several reasons:

First, it can be tailored to suit the level of a wide variety of students. You can start with $1 + 1 = 2$ and quickly work your way to the concept of phi or you can approach it from the angle of quadratic equations, not to mention even more complex formulations.

Secondly, you can look at it in art, in science, even in motor racing...

The problem was literature. At my school, we were keenly aware that examples of the golden ratio in literature were extremely thin on the ground and that, without the support of our colleagues in English, our proposal would be doomed to failure.

At a fateful moment, the work of Joan Helm was brought to my attention.

However, the prospect of developing resources to teach the average twelve year old medieval French poetry was far too daunting. I was prepared to give medieval English poetry a shot, providing I could find sufficient poems with a suitable level of mathematics for analysis. It had to be simple enough for a pre–teen of very average ability to tackle. But not too simple. It needed a degree of complexity but nothing overly difficult.

Somewhere in the dim recesses of my memory, I recalled that JRR Tolkien had translated *Sir Gawain and the Green Knight* and that, if I remembered it correctly, there was a scene about a pentangle. Surely if the golden ratio were a design element anywhere in the work, it would be there. I retrieved the volume from my shelves and realised at once it didn't show any evidence of such design. However, the description of the pentangle began on Line 619.

It was suspiciously close to the defining digits of the golden ratio. And that shouldn't have been possible. According to the received history of mathematics, the introduction of the decimal system to Europe was still centuries off.

By the time the work of Ed Condren had come to my attention, I was not only very suspicious on a number of fronts but, in scouting through many poetry collections of the era, I found some excellent material for my students to work with. So many medieval poems showed interesting and irregular features.

I was thrilled very early on to find many examples of multiples of 11 because, in teaching number patterns, it's difficult to keep the top students engaged while showing the lower half—yet again—how to recognise a multiple of 9 or even 10 or 5. Multiples of 11 were extremely simple to recognise but even the most advanced student had not encountered the process of uncovering them before. So they were a bonus.

At first it didn't occur to me that there was any special reason for these multiples. It was only in coming to the end of Condren's book on the *Pearl* manuscript it dawned on me there was an obvious significance to the number 490. That it could be, and almost certainly was, a metaphor for forgiveness.

It was a short step to the Grail question and asking myself what all those multiples of 11 meant. Or rather, as I quickly realised, what those multiples of 22 meant.

If 490 was a metaphor for forgiveness, what was the significance of 22?

And, even before it, what could 101 possibly mean? It was

quickly obvious this was the unanswered puzzle in so much of the literature on Cotton Nero *A.x*. Were those two sets of 101 stanzas just a coincidence or did they offer a significant clue, as many writers suggested, towards identifying the *Pearl* poet and the *Gawain* poet as the same person?

From reading Condren's work, I felt it was obvious where the answer would be found. In astronomy.

Condren had made much of the quadrivium in his analysis, looking at arithmetic, geometry and music. However the fourth member of the medieval 'integrated curriculum', astronomy, was notably absent. So I started my search there.

It took all of five minutes to find an ancient astronomical concept for 101. It happened to be related to music as well. The diatonic comma—the Pythagorean comma—was intimately linked to the notion of Music of the Spheres. Condren's suggestion that a musical design was a prime motivation for *Pearl* poet's numerical choices seemed further vindicated at that point.

From this discovery on, it was a simple matter of factorising every single major structural number in sight and asking what possible metaphor each of them might indicate. When factors produced no result, I'd apply the golden ratio or just simply add some metaphor I'd already discovered in case it yielded something positive.

Very quickly, my initial reasons for looking at the text were superseded. The quest was on as I became far more interested in the thrill of discovery than any possible pedagogical use. However I've told this story to make a very particular point: with only a few exceptions, I would expect a twelve year old of average mathematical ability to be able to recognise the arithmetic patterns in the *Pearl* manuscript.

This is where it comes down to mindset. Far too many criticisms of numerical literary form seem to suggest it's too hard to either produce or analyse for all but the most mathematically adept.

This is not the case at all.

Admittedly however many writers make it look hard. It took until the very last page of Condren's *The Numerical Universe of the Gawain-*

Pearl Poet: Beyond Phi to convince me he was right—partly because 490 was such an obvious metaphor and partly because I realised that repeated applications of the golden ratio would result in many of the same numbers as he found. A much simpler methodology would bring out the same design features. This was a quite a relief to me because I did not believe the way humans mentally strategise when it comes to number calculation would have changed much over the centuries, however much our philosophy about numbers might have altered.

As it turned out, I was extremely fortunate to have started with the work of Joan Helm. I was therefore conscious from the outset of the hidden numbers she had found significant in various works of Chrétien de Troyes. To find these same numbers as the most blatant construction devices in the *Pearl* manuscript begged for an explanation, particularly given the fact the *Pearl* poet did not appear to be Chrétien's biggest fan. However when the number 17 began to emerge as a major motif, the theological direction of the manuscript became abundantly clear.

The major difficulties I encountered during this analysis almost always occurred because so many 'scribal errors' have been amended by enthusiastic editors. Once I'd established a defined number pattern within the text, I'd follow a standard mathematical procedure for checking whether or not I had indeed discovered the poet's equation of choice. Time and again, I'd dismiss a particular hypothesis I'd formulated about a specific number I suspected was a metaphor. I knew that, if I were right about its metaphoric nature, I should be able to find some sort of highlight on an exact line. It was only much later an obscure mention of a 'scribal error' in the original manuscript which had been amended by an editor would come to light.

In rectifying these 'mistakes', editors have laid aside much evidence for numerical design. They have imposed their own beliefs about regularity on the text, rather than let it speak for itself. The mathematical construction in the original is very evident—however it has effectively vanished in many modern editions because its nature has not been understood.

The poetic conventions of the nineteenth, twentieth and

twenty–first centuries have dictated a philosophic view of the poetic conventions of the medieval era that is, quite simply, completely wrong.

This remark should alert everyone to the fact I've never abandoned that antiquated mathematical concept of a 'right' and 'wrong' answer, despite the efforts in some pedagogical quarters to class this as bigotry and establish a philosophy of 'mathematical consensus'. If such a worldview were ever to take serious hold, then its relativism will make the world a fearful place. I wouldn't want to step onto a plane or behind the wheel of a car designed by an engineer who believes in mathematical consensus. Such relativism would put paid to sentiments like those of Stephen Hawking who points out how remarkable it is that mathematics, as it currently stands, is actually able to predict the physical behaviour of objects. If mathematics is a human invention, then why does it work so well when it comes to describing a rocket trajectory or the flight of a ball through the air? 'What is it,' he has famously asked, 'that breathes fire into the equations and makes a universe for them to describe?'

The fire in the equations: it's such a tempting way to look at the *Pearl* poet's achievement. There is no disconnection in his view of the universe—his equations involve as many variables (metaphors) as possible.

His is a truly astonishing feat.

The fragmentation of modern studies into ever finer specialties has influenced our views of the past, seeing them as equally disjointed.

To choose an example from my own specialty: measurement in the modern era is so detached from everyday experience and tangibility that a metre is defined as *1650763.73 wavelengths of the red–orange emission line in the electromagnetic spectrum of the krypton–86 atom*. However for our forebears measures and measurement were inextricably linked to our own humanity. For the classical author, the human body was the template for 'due measure' while for the author of the Advent lyrics and the *Pearl* poet after him, the body of Christ was the supreme revelation of 'God the geometer'.

As we look back, it is evident that, for more than two centuries, a poetic war was fought over the meaning of measurement as it pertained to the ancient notion of the indissoluble union of words and number. Unless we recognise this, we fail to understand most of what the poets of old were saying. Statistical analysis of literary work is, at best, a doubtful tool. But simple mathematics—arithmetic, geometry, number—can reveal a style so distinctive that it is as good as a signature.

The *Pearl* poet's work is almost certainly the zenith of at least two millennia of numerical literary writing. It was also the fall of night for mathematical metaphor. It disappearance after such long and widespread acceptance begs so many questions.

Why did it vanish?

Perhaps it was simply that the heart was plucked out of it. As we have seen, the core of numerical literary form is the alignment of the golden ratio with the concept of *trawthe*. Troth: truth, faith, fidelity, ultimate reality. Whether a poet sympathised with neo–Platonic ideals and pointed to Delphi or whether, on the other hand, he was a zealous Christian and emphasised the Cross, the argument was always about that famous question of Pontius Pilate: 'What is truth?'

However, at the time of the Reformation, religious truth became a dangerously risky commodity for poets to commit themselves to in any public fashion. Within a generation of Henry VIII's proclamation of himself as the head of the Church in England and the dissolution of the monasteries, the crown's expectation of religious loyalty had violently see–sawed back and forth several times between Catholic and Protestant.

Poets who wanted to preserve their lives would have had to have become adept at hiding their own personal political and religious persuasions behind slippery wordings and double and triple entendres. In such a volatile era, truth would have had to have been the first casualty. The concept of arithmetic metaphor associated with truth naturally withered with it and, over time, the art of mathematical grammar was all but forgotten.

It is perhaps no coincidence that Milton, unafraid to proclaim his own religious and political loyalty even when his own life was at risk—he believed his life was forfeit when the restoration of Charles II was imminent—still remembered it and seems to have used it in *Paradise Lost*. Indeed, he used 101 as well and, given his life–long interest in the Music of the Spheres, this is probably no coincidence.[354]

As I neared the end of this investigation and discovered the work of Maren Sofie Røstvig on the arithmology of various Renaissance poets, a curious confirmation came to light.[355] She notes in her opening chapter to *The Hidden Sense* that, on the basis of the evidence presented by Curtius and Huisman,[356] 'the "mystical art of writing by number" was a well–known compositional technique in the Middle Ages, and that it embraced the counting of single letters, syllables, words, stresses, lines, stanzas, rhymes, groups of lines in the form of invocations or epic speeches or verse paragraphs, chapters, books, Cantos and Canticles. It was applied to prose as well as poetry, to brief lyrics and to epics, the point being that the symbolism inherent in the chosen numbers should be closely related to the contents.'[357]

It seems the work of Curtius and Huisman has never been translated into English and consequently remains unknown. However Røstvig's description of it is very much in accord with the research of David Howlett into early British literature and totally congruent with the findings of Joost Smit Sibinga and Maarten Menken.

In the centuries before the divorce of arithmetic and literature and their fragmentation into ever smaller specialties, there was an age when academic integration was a reality, not a distant ideal. It is an era on which we can only look back in wonder.

Although we will probably never see the like of the *Pearl* cycle again, at least we can gaze on it with a sense of admiration.[358] Once, long ago, a poet had a splendid vision of mathematical and verbal unity that ranged across the cosmos. He was probably not the first to combine the numbers and ideas he did. Far too many of the numbers emphasised by the *Pearl* poet turn up in David Howlett's analysis of the Gnomic poems of the Exeter Book.[359] Many more of them are found, according to Sibinga and Menken, in the structural design of the Christian gospels and epistles.

There's a difference, however. The *Pearl* poet had both the mathematical dexterity and the talent with both word and allusion to bring the vision to fruition in a single cycle of poems. Linked and locked, woven and sealed, crafted and refined, his writing gives little hint of the careful engineering required to dovetail it together. Only very occasionally when one or two critics note that a digression seems overlong is there a remez of something more.

As I've looked across the centuries, I've occasionally found faint hints of writers who might have known of numerical literary style or re-invented something like it: DH Lawrence in *The Tortoise* seems very clear, Oliver Wendell Holmes in *The Chambered Nautilus* is a possibility. The poem seems reflect the broken nautilus Holmes found on the beach which inspired the poem: it appears to have a broken Fibonacci pattern. Joan Helm claims AD Hope knew of fusing words and number patterns.

In my own blog and novel-writing, I have taken to the idea in a minor way. Using the same theme as the *Pearl* poet—the kiss of heaven and earth—I have written a 77777-word children's fantasy, *Daystar*.

I certainly didn't even try to reproduce the rich layering of mathematics in the poems of *Cotton Nero A.x.* But I was curious as to just how hard it is to produce chapters whose word lengths correspond to, for instance, the golden ratio. As I devised the mathematical architecture for the chapters, I began to recognise some numbers as the same as those in *Cotton Nero A.x.* Even though I hadn't planned them to be so.

It was a thrill to discover some of the more obscure aspects of the *Pearl* poet's design this way. I also set out to make this book conform to a word count of 77777.

If imitation is the sincerest form of flattery, then such a choice seemed a suitable way to honour the achievement of the *Pearl* poet.

So, if you've sometimes considered a digression rambling and repetitive or a point just a little too protracted, there's a reason. I had a goal with my word count and my internal design, so sometimes one section might feel a bit clipped and others might seem too fat.

Like the gloss on the pentangle Blanch and Wasserman saw as

overlong and Tolkien thought was elaborate and pedantic, the length of this book may be unsuitable for modern taste. Certainly it doesn't keep to academic strictures—it doesn't stay within the confines of fourteenth century medieval poetry or even mathematics. It trespasses into theology and roams through music and astronomy. It talks about comets and alludes to nationalistic fervour.

Perhaps the *Pearl* poet wasn't interested in all these things—but I suspect he was. I suspect it's hard to understand him or his work from the outside—that it's necessary to walk in his shoes for a moment to have any hope of comprehending what he was trying to achieve.

To understand his work was meant to soar upward to span both heaven and earth.

In my view, his achievement is unparalleled. His debate of the Four Daughters of God is a work of effortless grace and monumental mathematical elegance.

Appendix 1

The *Pearl* Poet

A Summary

The *Pearl* poet was a man with a nimble mind who loved word games: puns and subtleties and double, even triple, entendres. He was a mathematician of extraordinary skill who created artforms of geometric and arithmetic patterns within his poems. Besides *Pearl*, *Purity*, *Patience* and *Sir Gawain and the Green Knight*, he also wrote *St. Erkenwald.* In addition, he is extremely likely to be the author of *Summer Sunday* and *Death and Liffe*.

He loathed neo–Platonism in any of its forms—courtly love, high magic, occult mathematics, Catharism, Pythagoreanism—but his especial target was the work of Chrétien de Troyes. It seems likely he knew Geoffrey Chaucer, since they both use very similar constructions (with the obvious exception of 17). If he didn't know him, he was acquainted with his work since his choice of design numbers seems to have been influenced by Chaucer, though in reactive sense. (On the other hand, it could be suggested that Chaucer was reacting to him.) He would clearly have disapproved of Chaucer's apparently slippery flirtation with neo–Platonic ideas.

His work shows a liking for the legend of Pope Gregory, an abhorrence of Delphi and admiration for Paul's Epistle to the Corinthians and the Ephesians. His favourite design numbers were

17, 101, 153, 231, 490, 1111 and the golden ratio. He used 1743 too, but not repeatedly. In common with his contemporaries, he used multiples of 22. He avoided the use of multiples of 10, perhaps associating them with the tetraktys of neo–Pythagoreanism. He understood the arithmetic and philosophy of the decimal system and was familiar with the numerical literary techniques of the New Testament. He was very theologically minded, knew both Greek and Hebrew, had access to the works of the French neo–Platonists, was almost certainly educated in a monastery and may have been a rector.

Circumstantial evidence suggests he was John de Massey (perhaps known as Hugh to distinguish him from another rector named John de Massey), that he knew a Jordan de Holme and that he lived at either Sale, Stockport or Ashton–on–Mersey, in the vicinity of Baggily Hall, Cheshire. He may have been poor. He was possibly a Lollard, but if not, he seems to have sympathised with their cause. He wrote religiously themed verse in the vernacular which is the probable reason his poems have survived in only one copy. However he was so intelligent that it seems highly unlikely he wrote only in the vernacular and only in a Midland dialect.

Somewhere, sometime, it's more than likely someone will find the same mathematical signature I have found in a suite of Latin poems. I only hope they have a name attached to them.

Appendix 2

Numbers in the Numerical Literary Architecture

including comparisons with other possible works of the *Pearl* poet

Abbreviations: D&L (*Death and Liffe*), Pa (*Patience*), Pe (*Pearl*), Pu (*Purity/Cleanness*), SGGK (*Sir Gawain and the Green Knight*), StE (*St. Erkenwald*), SS(F) (*Summer Sunday/Fortune*)

11 or **22**, multiples of; token of the Wheel of Fate or the Wheel of the Divine Potter –

- 2530 (number of lines in SGGK excluding last line in Old French)
- 858 lines sub–section within Pe
- 1111 lines in concatenated Pe
- 5555 lines in Pe + Pu + SGGK
- 555.5 lines into Pu as dourad of Pe + Pu + SGGK

17, multiples of; token of Christian allegiance –

- 51 (number of lines in the description of Gawain's shield in the Arming Sequence, also appears in SS(F))
- 153 (the 'resurrection number' – used as a link in both SGGK and StE)
- 459 (number of lines in D&L which mentions

climbing a mountain with a view of 17 miles in every direction)

- 969 (a tetrahedral number and the sum of the first 17 triangular numbers; number of lines between the *tau* of Pa and the mention of Golden Mass in SGGK)
- 4913 (17 × 17 × 17, the number of lines in the total manuscript when the structural 'zeroes' are excluded)
- 6086 (total number of lines in the Pe manuscript)

101, multiples of; token of Music of the Spheres –

- 101 (number of stanzas in Pe + SGGK, also number of concatenations in Pe and 'bob and wheels' in SGGK)
- 202 (number of lines between 'phares' and the false gods; number of lines from end of Gawain's ordeal to end SGGK; *tau* of Pa)
- 303 (number of lines in Day 3 of Fitt 3 of SGGK)
- 505 (number of lines in the 'bob and wheels' of SGGK
- 555.5 (line position of the dourad in Pu)
- 1010 ($1010 \times \tau^2 = 386$, the Divine Potter of Pe – though there is some evidence this line has been shifted to 1011 in order to make a difference of 153 and also possibly to avoid a multiple of 10)
- 1111 (number of lines in Pe with concatenations counted as single lines)
- 1212 (number of lines in Pe)
- 1812 ($101 \div \tau^6$ = number of lines in Pu)
- 2121 lines is the *paz* of 5555 lines and Line 909 of Pu which is a warning to Lot to leave immediately.
- 2525 lines to the end of the 'long' lines in SGGK,

noted by Kent Hieatt.

- 3434 lines, the *tau* of 5555 lines and Line 410 of SGGK, just before the Green Knight exclaims, 'But stop!'
- 4343 (number of lines in Pu + SGGK)
- 5151 (the 101st triangular number: at this line position 1 man + 100 hounds are mentioned)
- 5454 (the number of lines from the mention of phoenix in Pe to the end of Gawain's ordeal with another 'Stop!': both Resurrection motifs)
- 5555 (number of lines in Pe + Pu + SGGK)
- 5656 (the number of lines from the mention of phoenix in Pe to the end of manuscript)

(There are 17 examples above, but I suspect I have missed finding one, or possibly two, depending on whether the poet's count included 1812 and the 'long' lines of SGGK.)

153, structural number, token of resurrection (however, note the pattern of usage regarding the Five Wounds on the Cross) –

- number of lines from the end of Fitt I of SGGK to the mention of 'Five Wounds of Christ'
- number of lines in Pe from descent of heavenly Jerusalem to mention of the fifth wound of the Lamb
- number of lines in Pe from mention of beryl (gemstone emblematic of Five Wounds) fettled with multiple of 101 (emblematic of Music of Spheres) to a multiple of 22 (emblematic of the Potter's Wheel)
- number of lines in St.E to major anomalous verse. St. E. partitioning seems to reflect Five Wounds on the Cross.
- $^{153}/_{265}$ is the standard fractional approximation for $^{1}/_{\sqrt{3}}$ since the time of Archimedes. $^{153}/_{265}$ of the manuscript is 490 lines into Pa.

231, structural number, token of resurrection and of Trinity –

- number of lines in Pe from the jeweller spying the descent of New Jerusalem to end of poem
- golden section of Pe at 462 or 2 × 231
- number of lines from second division of StE to end
- number of lines in Pe from token of forgiveness (980 = 2 × 490) to end of poem
- arithmetic equivalent of geometric triple triangle and counterpoint to tetraktys

490, defining number, token of forgiveness –
number of lines in Fitt I of SGGK

- number of lines from Gawain tying on green girdle to the end of SGGK
- multiplied by τ = 303, number of lines in Day 3 of Fitt III of SGGK
- divide by 2 and subtract 144 = 101, mathematical token of Music of the Spheres
- multiplied by 10 = line numbering when all 'zeroes' except verb and plural forms of 'poynte' are considered
- likely sub–structure of Pa (490 + 40 + 1 = 531)
- Line 490 of Pu mentions forgiveness/reconciliation
- Line 490 of SGGK (size of Fitt I) is the dourad of Line 1540. 1540 is the 55th triangular number which makes it triply triangular.
- 490 lines from Line 618 of SGGK is Line 1108 which mentions 'swearing by trawthe'.
- 490 lines into Pa is $^{153}/_{265}$ of the manuscript and $^{153}/_{265}$ is the standard fractional approximation for $^{1}/_{\sqrt{3}}$ since the time of Archimedes: it is the beginning of a discourse by God about mercy and forgiveness.

111 *and* **1111**, structural numbers, tokens of heavenly city –

- 1111 is number of lines in Pe with concatenations counted as single lines
- 1111 multiplied by 5 = combined total of Pe, Pu and SGGK
- 1111 multiplied by 0.5 = line numbering in Pu of dourad position of Pe, Pu and SGGK
- 1111 is size of an internal wheel in Pu
- Line 1111 in Pe mentions God on his throne

618, structural number; token of trawthe/truth/faith/God and also peace –

- 0.618 :the golden section – used continually (at least 8 times) to refer to a Christian concept of truth vis à vis a neo–Platonic or Pythagorean concept of truth. Used to link the defining number, 490, to 101, the most frequently observed structural number.
- 0.382 (or 1 – 0.618): the golden section – used continually (at least 8 times) to refer to a Christian concept of truth vis à vis a neo–Platonic or Pythagorean concept of truth.
- 618 lines from beginning of SGGK to description of Gawain's shield
- 618 lines backwards from beginning of SGGK to Writing on the Wall
- 2 × 618 lines from Line 490 (mentioning forgiveness) of Pu to the Writing on the Wall
- 618 lines from end of Fitt I (490 lines) of SGGK is Line 1108 which mentions 'swearing by trawthe'.
- approximately 4 × 618 from one Daniel reference to next: Master of Spiritual Mysteries to Writing on the Wall.

- approximately 5 × 618 from Master of Spiritual Mysteries to beginning of SGGK.
- approximately 6 × 618 from Master of Spiritual Mysteries to description of Gawain's shield
- $1010 × \tau^2 = 386$, the Divine Potter of Pe
- $101 ÷ \tau^6 = 1812$, number of lines in Pu
- 2121 lines is the *paz* of 5555 lines and Line 909 of Pu which is a warning to Lot to leave immediately
- 3434 lines the *tau* of 5555 lines and Line 410 of SGGK, just before the Green Knight exclaims, 'But stop!'
- 62, the 'superfluity' or extra lines over the even thousand in the second 'half' of the manuscript: $100 × \tau$ to the nearest whole number
- 24, the 'superfluity' or extra lines over the even thousand in the first 'half' of the manuscript: $100 × \tau^3$ to the nearest whole number

Appendix 3

Meaning of Common Mathematical Metaphors

(1)   Truth: the Golden Section (either as 0.618 or as 0.382) or 231

(2)   The Wheel of Fate (multiples of 11 or 22)

(3)   The Kingdom of Heaven (arithmetically: multiples of 1111 or a ratio 1.111; geometrically: square within a circle)

(4)   Resurrection (153 or 231)

(5)   Terrestrial Geography (combinations of 3, 10, 12, 22, 40, 144 with 176)

(6)   Solar and–or Lunar Calendar (354, 29, 365, 366, 52)

(7)   Christianity (multiples of 17)

(8)   Forgiveness, mercy, reconciliation (490)

(9)   Incarnation (arithmetically: $1/_3$, $2/_3$, 0.82665; geometrically: equilateral triangle within a circle)

(10)   The Kiss of Earth and Moon or The Kiss of Heaven and Earth (1742 or 1743)

(11)   The Music of the Spheres (multiples of 101)

Appendix 4

The Pentangle Re-considered

Line 619

This is the start of the description of the pentangle. At just one line after the significant digits of the golden ratio in decimal notation, it highlights not only the use of that Hindu–Arabic system but also links the line numbering to the mathematics of the pentangle itself.

Line 621

1111 – 490 kiss/mercy

Line 624

777 – 153 armour/resurrection, also 12 × 52 peace? Time?

Line 632

1743 – 1111 kiss/ kiss

Line 638

Paz of Pentangle description and tau of part II
Intensified truth

Line 643

153+490resurrection/mercygoldenmass/resurrection/forgiveness

Line 644

Halfway – balance?

Line 651

540 + 111 phoenix/kiss

Line 652

805 – 153 wheel of fate/resurrection

Line 667

Pythagoras

Line 669

End of the sequence is 999 lines from the paz of Patience

Bibliography

Acheson, D., *1089 and All That – A Journey into Mathematics*, Oxford University Press, 2002

Ackermann, E.C., *The Golden Section*, The American Mathematical Monthly, Vol. 2, No. 9/10, Mathematical Association of America, 1895

Adams, H., *Mont Saint Michel and Chartres*, Princeton University Press 1933

Addis, D., *Sir Gawain and the Green Knight and Its Biblical References*, http://www.uh.edu/hti/cu/2005/7/01.pdf

Adolf, H., *A Historical Background for Chrétien's Perceval*, PMLA, Vol. 58, No. 3, Modern Language Association, 1943

Aers, D., *The Self Mourning: Reflections on Pearl*, Speculum, Vol. 68, No. 1, 1993

Aker, P., *The Emergence of an Arithmetical Mentality in Middle English Literature*, The Chaucer Review, Vol 28, No 3, 1994

Alder, K., *The Measure of All Things: The Seven-Year Odyssey that Transformed the World*, Little, Brown 2002

Alford, J.A., *The Grammatical Metaphor: A Survey of Its Use in the Middle Ages*, Speculum, Vol. 57, No. 4, Medieval Academy of America, 1982

Allen, V., *Sir Gawain: Cowardyse and the Fourth Pentad*, The Review of English Studies, New Series, Vol. 43, No. 170, 1992

Anderson, J.J. (tr.), *Patience*, Manchester University Press, 1969

Anderson, J.J., *Language and Imagination in the Gawain–Poems*, Manchester University Press, 2005

Andrew, M., Waldron, R., (eds.), *The Poems of the Pearl Manuscript: Pearl, Cleanness, Patience, Sir Gawain and the Green Knight*, Edward Arnold (Publishers) Ltd., 1978

Andrew, M., *The Diabolical Chapel: A Motif in Patience and Sir Gawain and the Green Knight*, Neophilologus, Vol. 66. No. 2, 1982

Armitage, Simon (tr.), *Sir Gawain and the Green Knight*, W.W. Norton & Co, 2007

Arthur, R.G., *Medieval Sign Theory and Sir Gawain and the Green Knight*, University of Toronto Press, 1987

Asher, D.J.; Bailey, M.; Emel'yanenko, V.; Napier, B., *Earth in the Cosmic Shooting Gallery*, Observatory 125, 2005

Aston, M.E., *Lollardy and Sedition 1381–1431*, Past and Present, No. 17, Oxford University Press, 1960

Atlas, A.W., *Gematria, Marriage Numbers, and Golden Sections in Dufay's "Resvellies vous"*, Acta Musicologica, Vol. 59, Fasc. 2, 1987

Austin, H.D., *Dante's Precious Stones and Those of the Heavenly City*, Italica, Vol. 22, No. 2, American Association of Teachers of Italian, 1945

Ayto, John, *Dictionary of Word Origins*, Bloomsbury, 2001

Babcock, C.F., *A Study of the Metrical Use of the Inflectional e in Middle English, with Particular Reference to Chaucer and Lydgate*, PMLA, Vol. 29, No. 1, 1914

Bachmann, T.; Bachmann, P.J., *An Analysis of Béla Bartók's Music through Fibonaccian Numbers and the Golden Mean*, The Musical Quarterly, Vol. 65, No. 1., 1979

Backhouse, J., *The Lindisfarne Gospels*, Phaidon Press 1981

Baillie, M., *Exodus to Arthur – Catastrophic Encounters with Comets*, B.T. Batsford Ltd 1999

Baillie, M., *New Light on the Black Death: The Cosmic Connection*, Tempus 2006

Baillie M., McCafferty, P., *The Celtic Gods: Comets in Irish Mythology*, Tempus 2005

Baily, J., *St Hugh's Church at Lincoln*, Architectural History, Vol. 34, SAHGB Publications Limited 1991

Baker, D., *The Gödel in Gawain – Paradoxes of Self-reference and the Problematics of Language in Sir Gawain and the Green Knight*, Cambridge Quarterly 2003

Bamford, C., *Homage to Pythagoras*, Lindisfarne Letter 14, 1982

Bannister, H. M., *The Introduction of the Cultus of St. Anne into the West*, The English Historical Review, Vol. 18, No. 69 Oxford University Press, 1903

Barber, R.; Barker, J., *Tournaments: Jousts, Chivalry and Pageants in the Middle Ages*, The Boydell Press, 2000

Barber, R., *The Holy Grail – Imagination and Belief*, Allen Lane 2004

Barfield, O., *History in English Words*, Lindisfarne 2000

Barraclough, J., *Systems of Exchange and Reciprocity in Sir Gawain and the Green Knight*, McGill University Montreal1989

Barrett, Jr, R.W., *Against All England: Regional Identity and Cheshire Writing, 1195–1656*, ReFormations, University of Notre Dame Press, 2009

Barron, W.R.J., *Trawthe and Treason: The Sin of Gawain Reconsidered*, Manchester University Press, 1980

Barron, W.R.J., *Chrétien and the Gawain–Poet: Master and Pupil or Twin Temperaments?* in Lacy, N., Kelly, D., Busby, K., *The Legacy of Chrétien de Troyes* Vol.II, Editions Rodopi, 1988

Bartelt, A.H., *The Book Around Immanuel – Style and Structure in Isaiah 2 – 12*, Eisenbrauns 1996

Bates, L.R., *Reading and Believing: Covenant in the Poems of the Pearl Manuscript*, Medieval Reading Group at the University of Cambridge

Bax, C., *Vintage Verse: An Anthology of Poetry in English*, Hollis and Carter Ltd., 1945

Bazak, J., *Numerical Devices in Biblical Poetry*, Vetus Testamentum, Vol. 38, Fasc. 3, 1988

Beck, J., *Piero della Francesca at San Francesco in Arezzo: An Art–Historical Peregrination*, Artibus et Historiae, Vol. 24, No. 47, IRSA s.c., 2003

Bedos–Rezak, B.M., *Medieval Identity: A Sign and a Concept*, The American Historical Review, Vol. 105, No. 5, American Historical Association, 2000

Benson, L.D., *The Authorship of St. Erkenwald*, Journal of English and Germanic Philology 1965

Benson, L.D., *The Style of Sir Gawain [Critical Studies of Sir Gawain and the Green Knight*, ed. Donald R. Howard and Christian K. Zacker] University of Notre Dame Press 1970

Benson, L.D., *The Meaning of Sir Gawain and the Green Knight* [*Critical Studies of Sir Gawain and the Green Knight*, ed. Donald R. Howard and Christian K. Zacker] University of Notre Dame Press 1970

Bercovitch, S., *Romance and Anti-Romance in Sir Gawain and the Green Knight* [*Critical Studies of Sir Gawain and the Green Knight*, ed. Donald R. Howard and Christian K. Zacker] University of Notre Dame Press 1970

Bergner, H., *Sir Orfeo and the Sacred Bonds of Matrimony*, The Review of English Studies, New Series, Vol. 30, No. 120, Oxford University Press 1979

Besserman, L., *The Idea of the Green Knight*, *ELH*, Vol. 53, No. 2. Summer, 1986

Bishop, I., *The Significance of the 'Garlande Gay' in the Allegory of Pearl*, The Review of English Studies, New Series, Vol. 8, No. 29, Oxford University Press 1957

Bishop, I., *Time and Tempo in Sir Gawain and the Green Knight*, Neophilologus, Vol. 69, No. 4, 1985

Bishop, L. M., *Words, Stones and Herbs – The Healing Word in Medieval and Early Modern England*, Syracuse University Press, 2007

Blanch, R.J.; Wasserman, J.N., *From Pearl to Gawain – Forme to Fynisment*, University Press of Florida 1995

Blenkinsopp, J., *Structure and Style in Judges 13–16*, Journal of Biblical Literature, Vol. 82, No. 1, The Society of Biblical Literature, 1963

Bloomfield, M.W., *Sir Gawain and the Green Knight: An Appraisal* [*Critical Studies of Sir Gawain and the Green Knight*, ed. Donald R. Howard and Christian K. Zacker] University of Notre Dame Press 1970

Bolton, W.F. (ed), *The Middle Ages – Volume 1 of the Penguin History of Literature*, Penguin, 1993

Borroff, M., *Criticism of Style: The Narrator in the Challenge Episode* [*Critical Studies of Sir Gawain and the Green Knight*, ed. Donald R. Howard and Christian K. Zacker] University of Notre Dame Press 1970

Bos, A.P., *'Aristotelian' and 'Platonic' Dualism in Hellenistic and Early Christian Philosophy and in Gnosticism*, Vigiliae Christianae, Vol. 56, No. 3, BRILL 2002

Bowers, JM, *Chaste Marriage: Fashion and Texts at the Court of Richard II*, Pacific Coast Philology vol. 30. no.1., Pacific Ancient and Modern Language Association, 1995

Bowers, JM, *The Politics of "Pearl": Court Poetry in the Age of Richard II*, Boydell and Brewer, 2001

Bowers, RH, *Gawain and the Green Knight as Entertainment* [*Critical Studies of Sir Gawain and the Green Knight*, ed. Donald R. Howard and Christian K. Zacker] University of Notre Dame Press 1970

Braddy, H., *Sir Gawain and Ralph Holmes the Green Knight*, Modern Language Notes, Vol. 67, No. 4., Apr., 1952

Brewer, D., Gibson, J., *A Companion to the Gawain–Poet*. Ed.. Cambridge: D.S. Brewer, 1997

Briggs, K., *A Dictionary of Fairies, Hobgoblins, Brownies, Bogies and Other Supernatural Creatures*, Penguin Books 1979

Bucher, F., *Medieval Architectural Design Methods 800 –1560*, Gesta, vol.11 no. 2, International Center of Medieval Art 1972

Bulatkin, E.W., *Structural Arithmetic Metaphor in the Oxford "Roland"*, Ohio State University Press, 1972

Burnley, J.D., *Fine Amor: Its Meaning and Context*, The Review of English Studies, New Series, Vol. 31, No. 122, Oxford University Press 1980

Burns, M., *J. R. R. Tolkien: The British and the Norse in Tension*, Pacific Coast Philology, Vol. 25, No. 1/2, Pacific Ancient and Modern Language Association 1990

Burrow, JA, *A Reading of Sir Gawain and the Green Knight*, Routledge and Kegan Paul, 1965

Burrow, J., *The Two Confession Scenes in "Sir Gawain and the Green Knight"*, Modern Philology, Vol. 57, No. 2. Nov., 1959

Burrow, J., *"Cupiditas" in Sir Gawain and the Green Knight: A Reply to D.F. Hills* [*Critical Studies of Sir Gawain and the Green Knight*, ed. Donald R. Howard and Christian K. Zacker] University of Notre Dame Press 1970

Busi G, *Hebrew to Latin, Latin to Hebrew: the mirroring of two cultures in the age of humanism*: colloqium held at the Warburg Institute, London, October 18–19, 2004, Freie Universität Berlin. Institut für Judaistik

Cadden, J., *Science and Rhetoric in the Middle Ages: The Natural Philosophy of William of Conches*, Journal of the History of Ideas, Vol. 56, No. 1, University of Pennsylvania Press 1995

Campbell, J.J., *Structural Patterns in the Old English Advent Lyrics*, ELH, Vol. 23, No. 4. , 1956

Cargill, O.; Schlauch, M., *The Pearl and Its Jeweler*, PMLA, Vol. 43, No. 1, Modern Language Association 1928

Carson, A., *The Green Chapel: Its Meaning and Its Function* [*Critical Studies of Sir Gawain and the Green Knight*, ed. Donald R. Howard and Christian K. Zacker] University of Notre Dame Press 1970

Caspar, D. L. D.; Fontano, E., *Five-Fold Symmetry in Crystalline Quasicrystal Lattices*, Proceedings of the National Academy of Sciences of the United States of America, Vol. 93, No. 25, National Academy of Sciences 1996

Casti, J., *I Know What You'll Do Next Summer*, New Scientist (August 2002)

Celenza, C. S., *Pythagoras in the Renaissance: The Case of Marsilio Ficino*, Renaissance Quarterly, Vol. 52, No. 3, Renaissance Society of America 1999

Chapman, C.O., *The Musical Training of the Pearl Poet*, PMLA, Vol. 46, No. 1. Modern Language Association 1931

Chapman, C.O., *The Authorship of the Pearl*, PMLA, Vol. 47, No. 2. Modern Language Association 1932

Chapman, C.O., *Virgil and the Gawain-Poet*, PMLA, Vol. 60, No. 1. Modern Language Association 1945

Chapman, C.O., *Ticius to Tuskan, GGK, Line 11*, Modern Language Notes, Vol. 63, No. 1, The Johns Hopkins University Press 1948

Chapman, C.O., *Index of Names in Pearl * Purity * Patience * Gawain*, Cornell University Press 1951

Chapman, C.O., *Chaucer and the Gawain-Poet: A Conjecture,* Modern Language Notes, Vol. 68, No. 8., 1953

Chaucer, G.; Brewer, D.S. (ed.), *The Parlement of Foulys*, Thomas Nelson and Sons Ltd., 1960

Chaucer, G.; Coghill, N (tr.), *Troilus and Criseyde*, Penguin Books, 1971

Cherewatuk, K., Echoes of the Knighting Ceremony in *Sir Gawain and the Green Knight*, Neophilologus, Vol. 77., No.1, 1993

Cherry, J., *Late Fourteenth-Century Jewellery: The Inventory of November 1399*, The Burlington Magazine, Vol. 130, No. 1019, Special Issue on English Gothic Art, The Burlington Magazine Publications, Ltd. 1988

Chrétien de Troyes; Gilbert, D. (tr.), *Erec and Enide*, University of California Press,1992

Christmas, P., A Reading of *Sir Gawain and the Green Knight*, Neophilologus, Vol. 58, No. 2, 1974

Clark, J.W., *Paraphrases for "God" in the Poems Attributed to "The Gawain-Poet"*, Modern Language Notes, Vol. 65, No. 4. Apr., 1950

Clark, S. L.; Wasserman, J.N., *The Passing of the Seasons and the Apocalyptic in "Sir Gawain and the Green Knight"*, South Central Review, Vol. 3, No. 1. 1986

Clason, R. G., *A Family of Golden Triangle Tile Patterns*, The Mathematical Gazette, Vol. 78, No. 482, The Mathematical Association 1994

Clemoes, P., *Rhythm and Cosmic Order in Old English Literature*, Cambridge University Press, 1970

Clermont-Ganneau, M.; Rogers, R.W. (tr.), *Mene, Tekel, Peres, and the Feast of Belshazzar*, Hebraica, Vol. 3, No. 2., 1887

Clopper, L.M., *The God of the "Gawain-Poet"*, Modern Philology, Vol. 94, No. 1, The University of Chicago Press 1996

Cockcroft, R., *Castle Hautdesert: Portrait or Patchwork?*, Neophilologus, Vol. 62, No. 3, 1978

Condren, E.I., *The Paradox of Chrétien's Lancelot, MLN*, Vol. 85, No. 4, French Issue., 1970

Condren, E.I., *Chaucer and the Energy of Creation: The Design and Organization of the Canterbury Tales* University Press of Florida, 1999

Condren, E.I., *The Numerical Universe of the Gawain-Pearl Poet: Beyond Phi* University Press of Florida, 2002

Condren, E.I., *Chaucer from Prentice to Poet: The Metaphor of Love in Dream Visions and Troilus and Criseyde* University Press of Florida, 2008

Conway, D.J., *The Celtic Book of Names – Traditional Names from Ireland, Scotland and Wales*, Citadel Press, 1999

Cook, T. A., *The Curves of Life*, Dover Books, 1979

Cooke, J., *The Lady's 'Blushing' Ring in Sir Gawain and the Green Knight*, The Review of English Studies Vol. 49, No. 193 1998

Coomaraswamy, A.K., *Sir Gawain and the Green Knight: Indra and Namuci*, Speculum, Vol. 19, No. 1. Jan., 1944

Cooper, R. A.; Pearsall, D. A., *The Gawain Poems: A Statistical Approach to the Question of Common Authorship*, The Review of English Studies, New Series, Vol. 39, No. 155. Aug., 1988

Cornford, F. M., *Mathematics and Dialectic in the Republic VI.–VII. (II.)*, Mind, New Series, Vol. 41, No. 162, Oxford University Press 1932

Crawford, A., *The Queen's Council in the Middle Ages*, The English Historical Review, Vol. 116, No. 469 Oxford University Press 2001

Crombie, A.C., *Science, Art and Nature in Medieval and Modern Thought*, Hambledon Press, 1996

Cronan, D., *Poetic words, conservatism and the dating of Old English poetry*, Beowulf Studies 2006

Cronin, H. S., *The Twelve Conclusions of the Lollards*, The English Historical Review, Vol. 22, No. 86 Oxford University Press 1907

Curley, M. J., *A Note on Bertilak's Beard*, Modern Philology, Vol. 73, No. 1, The University of Chicago Press 1975

Curnow, T., *Oracles of the Ancient World – A Comprehensive Guide*, Duckworth 2004

d'Ardenne, S. R. T. O., *'The Green Count' and Sir Gawain and the Green Knight*, The Review of English Studies, New Series, Vol. 10, No. 38, Oxford University Press 1959

Darr, K.P., *The Wall around Paradise: Ezekielian Ideas about the Future*, Vetus Testamentum, Vol. 37, Fasc. 3, BRILL 1987

Davenport, A.A., *The Catholics, the Cathars, and the Concept of Infinity in the Thirteenth Century*, Isis, Vol. 88, No. 2, The University of Chicago Press 1997

Davenport, G., *The Geography of the Imagination*, Nonpariel 1997

Davis, N.M., *Gawain's Rationalist Pentangle*, ARTHURIAN LITERATURE XII, D.S. Brewer, 1993

Davies, E.W., *A Mathematical Conundrum: The Problem of the Large Numbers in Numbers I and XXVI*, Vetus Testamentum, Vol. 45, Fasc. 4, BRILL 1995

Davies, P., *The Goldilocks Enigma*, Penguin Books 2007

Davies, R.T., *Chaucer: The Prologue to the Canterbury Tales*, Harrap 1955

Davis, M.T., Neagley, L.E., *Mechanics and Meaning: Plan Design at Saint–Urbain, Troyes and Saint–Ouen, Rouen*, Gesta, Vol. 39, No. 2, *Robert Branner and the Gothic*, International Center of Medieval Art 2000

Davis, N., 1995, *Recognition of Worth in Pearl and Sir Gawain and the Green Knight* in *The Middle Ages in the North–West*, ed. Tom Scott and Pat Starkey, Leopard's Head Press, Liverpool

Deakin, M. A. B.; Lausch, H., *The Bible and Pi*, The Mathematical Gazette, Vol. 82, No. 494, The Mathematical Association 1998

Delaney, J.J., *Dictionary of Saints*, Doubleday 1980

Delany, P., *The Role of the Guide in Sir Gawain and the Green Knight* [*Critical Studies of Sir Gawain and the Green Knight*, ed. Donald R. Howard and Christian K. Zacker] University of Notre Dame Press 1970

Deming, W., *The Unity of 1 Corinthians 5–6*, Journal of Biblical Literature, Vol. 115, No. 2, The Society of Biblical Literature 1996

Derrickson, A., *The Pentangle: Guiding Star for the Gawain Poet*, Comitatus: A Journal of Medieval and Renaissance Studies: Vol. 11, 1980

Dewdney, A.K., *A Mathematical Mystery Tour – Discovering the Truth and Beauty of the Cosmos*, John Wiley and Sons, 1999

Dixon, R., *The Mathematics and Computer Graphics of Spirals in Plants*, Leonardo, Vol. 16, No. 2, The MIT Press 1983

Doczi, G., *The Power of Limits – Proportional Harmonies in Nature, Art and Architecture*, Shambala Publications, Inc 1981

Dodds, E. R., *The Parmenides of Plato and the Origin of the Neoplatonic 'One'*, The Classical Quarterly, Vol. 22, No. 3/4, Cambridge University Press 1928

Dow, H. J., *The Rose-Window*, Journal of the Warburg and Courtauld Institutes, Vol. 20, No. 3/4, The Warburg Institute 1957

Duckworth, G.E., *Structural Patterns and Proportions in Virgil's Aeneid*, The University of Michigan Press, 1962.

Duckworth, G.E., *Five Centuries of Latin Hexameter Poetry: Silver Age and Late Empire*, Transactions and Proceedings of the American Philological Association, Vol. 98, The Johns Hopkins University Press 1967

Duggan, H.N., *Strophic Patterns in Middle English Alliterative Poetry*, Modern Philology, Vol. 74, No. 3, The University of Chicago Press 1977

Duggan, H.N., *Final "-e" and the Rhythmic Structure of the B-Verse in Middle English Alliterative Poetry*, Modern Philology, Vol. 86, No. 2, The University of Chicago Press 1988

du Santoy, M., *The Music of the Primes: Why an Unsolved Problem in Mathematics Matters*, Fourth Estate, 2003

Eadie, J., *Sir Gawain's Travels in North Wales*, The Review Of English Studies, New Series, Vol. 34, No. 134, Oxford University Press 1983

Eadie, J., *Morgain La Fée and the Conclusion of Sir Gawain and the Green Knight*, Neophilologus Vol. 52, No. 1, 1968

Edgeworth, R.J.,*Anatomical Geography In Sir Gawain And The Green Knight*, Neophilologus, Vol. 69, No. 2, 1985

Edwards, M (tr.), *Neoplatonic Saints: The Lives of Plotinus and Proclus by their Students*, Liverpool University Press, 2000

Edwards, R.R., *Ratio and Invention: A Study of Medieval Lyric and Narrative*, Vanderbilt University Press, 1989

Eggebroten, A, *Sawles Warde: A Retelling of De Anima for a Female Audience*, Mediaevalia, Vol. 10, 1984

Emerson, O.F., *A Parallel Between The Middle English Poem Patience And An Early Latin Poem Attributed To Tertullian*, Publications of the Modern Language Association of America 10 (New series, vol. 3), 1895

Emerson, O.F., *Middle English*, Modern Language Notes, Vol. 28, No. 6, The Johns Hopkins University Press 1913

Emerson, O.F., *Middle English Clannesse*, PMLA, Vol. 34, No. 3, Modern Language Association 1919

Emerson, O.F., *Imperfect Lines in "Pearl" and the Rimed Parts of "Sir Gawain and the Green Knight"*, Modern Philology, Vol. 19, No. 2. Nov., 1921

Emerson, O.F., *Some Notes on the Pearl*, PMLA, Vol. 37, No. 1, (Modern Language Association 1922

Erickson, B., *Art and Geometry: Proportioning Devices in Pictorial Composition*, Leonardo, Vol. 19, No. 3, The MIT Press 1986

Fell, C. (tr.), Lucas, J., *Egils Saga*, J.M. Dent & Sons Ltd., 1975

Fein, S.G., *Twelve-Line Stanza Forms in Middle English and the Date of Pearl*, Speculum, Vol. 72, No. 2, Medieval Academy of America 1997

Fowler, D.C., *Cruxes in "Cleanness"*, Modern Philology, Vol. 70, No. 4, The University of Chicago Press 1973

Fischler, R., *On the Application of the Golden Ratio in the Visual Arts*, Leonardo, Vol. 14, No. 1. Winter, 1981

Frantzen, A.J., *The Disclosure of Sodomy in Cleanness*, PMLA, Vol. 111, No. 3., 1996

Franzen, C., *Sir Gawain and the Green Knight*, Victoria University of Wellington 1983

Freeman, P., *Proportion in Gothic Architecture – A Paper*, The Cambridge Architectural Society, 1848

Frugoni, C.; McCuaig. W. (tr.), *Inventions of the Middle Ages*, The Folio Society, 2007

Gandz, S., *On the Origin of the Term "Root"*, The American Mathematical Monthly, Vol. 33, No. 5, Mathematical Association of America 1926

Gandz, S., *On the Origin of the Term "Root." Second Article*, The American Mathematical Monthly, Vol. 35, No. 2, Mathematical Association of America 1928

Gandz, S., *The Origin of Angle-Geometry*, Isis, Vol. 12, No. 3, The University of Chicago Press 1929

Gandz, S., *The Knot in Hebrew Literature, or from the Knot to the Alphabet*, Isis, Vol. 14, No. 1, The University of Chicago Press 1930

Gandz, S., *Mene Mene Tekel Upharsin, a Chapter in Babylonian Mathematics*, Isis, Vol. 26, No. 1, The University of Chicago Press 1936

Gandz, S., Sarton, G., *The Invention of the Decimal Fractions and the Application of the Exponential Calculus by Immanuel Bonfils of Tarascon (c. 1350)*, Isis, Vol. 25, No. 1, The University of Chicago Press 1936

Gandz, S., *The Babylonian Tables of Reciprocals*, Isis, Vol. 25, No. 2, The University of Chicago Press 1936

Gandz, S., *The Origin and Development of the Quadratic Equations in Babylonian, Greek, and Early Arabic Algebra*, Osiris, Vol. 3, The University of Chicago Press 1937

Gandz, S., *Studies in Babylonian Mathematics II. Conflicting Interpretations of Babylonian Mathematics*, Isis, Vol. 31, No. 2, The University of Chicago Press 1940

Gandz, S., *Studies in Babylonian Mathematics III: Isoperimetric Problems and the Origin of the Quadratic Equations*, Isis, Vol. 32, No. 1, The University of Chicago Press 1940

Gandz, S., *Studies in Babylonian Mathematics I: Indeterminate Analysis in Babylonian Mathematics*, Osiris, Vol. 8, (The University of Chicago Press 1948

Gandz, S., *Studies in the Hebrew Calendar: I. A Study in Terminology*, The Jewish Quarterly Review, New Series, Vol. 39, No. 3, University of Pennsylvania Press 1949

Gandz, S., *Studies in the Hebrew Calendar: II. The Origin of the Two New Moon Days*, The Jewish Quarterly Review, New Series, Vol. 40, No. 2, University of Pennsylvania Press 1949

Gandz, S., *Studies in the Hebrew Calendar (Continued)*,The Jewish Quarterly Review, New Series, Vol. 40, No. 3, University of Pennsylvania Press 1950

Gandz, S., *The Calendar of the "Seder Olam"*, The Jewish Quarterly Review, New Series, Vol. 43, No. 2, University of Pennsylvania Press 1952

Gandz, S., *The Benediction over the Luminaries and the Stars*, The Jewish Quarterly Review, New Series, Vol. 44, No. 4, University of Pennsylvania Press 1954

Ganim, J. M., *Disorientation, Style, and Consciousness in Sir Gawain and the Green Knight*, PMLA, Vol. 91, No. 3, Modern Language Association 1976

Gardner, J., *The Alliterative Morte Arthure, The Owl and the Nightingale and Five Other Middle English Poems in a Modernized Version with Comments on the Poems and Notes*, Southern Illinois University Press, 1971

Garner, A., *Benighted Verse*, The Times, December 27, 2003

Gatta, Jr., J., *Transformation Symbolism and the Liturgy of the Mass in "Pearl"*, Modern Philology, Vol. 71, No. 3. Feb., 1974

Geoffrey of Monmouth; Thorpe, L. (tr), *The History of the Kings of Britain*, Penguin 1966

Ghyka, M., *Gothic Canons of Architecture*, The Burlington Magazine for Connoisseurs, Vol. 86, No. 504, The Burlington Magazine Publications, Ltd. 1945

Godwin, J., *Pythagoreans, Today?*, Lindisfarne Letter 14, 1982

Goldhurst, W., *The Green and the Gold: The Major Theme of Gawain and the Green Knight*, College English, Vol. 20, No. 2., Nov., 1958

Goodall, C., *Eigenshape Analysis of a Cut–Grow Mapping for Triangles, and its Application to Phyllotaxis in Plants*, SIAM Journal on Applied Mathematics, Vol. 51, No. 3, Society for Industrial and Applied Mathematics 1991

Grant, R. M., *The Resurrection of the Body*, The Journal of Religion, Vol. 28, No. 2, The University of Chicago Press 1948

Grant, R. M., *The Resurrection of the Body [continued]*, The Journal of Religion, Vol. 28, No. 3, The University of Chicago Press 1948

Grant, R. M., *"One Hundred Fifty–Three Large Fish" (John 21:11)*, The Harvard Theological Review, Vol. 42, No. 4, Cambridge University Press 1949

Grant, R. M., *The Earliest Christian Gnosticism*, Church History, Vol. 22, No. 2, Cambridge University Press 1953

Grant, R. M., *Dietary Laws among Pythagoreans, Jews, and Christians*, The Harvard Theological Review, Vol. 73, No. 1/2, Dedicated to the Centennial of the Society of Biblical Literature, Cambridge University Press 1980

Graves, R., *The White Goddess – a historical grammar of poetic myth*, Farrar, Strauss & Giroux 1999

Graves, R.N., *The Hidden Runic Poem in the Medieval Allegory Pearl*, http://www.utm.edu/staff/ngraves/shakespeare/PearlRune.htm (accessed January 2008)

Green, R.F., *Jack Philipot, John of Gaunt, and a Poem of 1380*, Speculum, Vol. 66, No. 2 Medieval Academy of America 1991

Green, R.H., *Gawain's Shield and the Quest for Perfection*, ELH, Vol. 29, No. 2, The Johns Hopkins University Press 1962

Greene, W.K., *The Pearl – A New Interpretation*, PMLA, Vol. 40, No. 4, Modern Language Association 1925

Greenberg, M., *Idealism and Practicality in Numbers 35:4–5 and Ezekiel 48*, Journal of the American Oriental Society, Vol. 88, No. 1, American Oriental Society 1968

Greenwood, O. (tr.), *Sir Gawain and the Green Knight*, Lion and Unicorn Press, 1956

Gross, K., *'Each Heav'nly Close': Mythologies and Metrics in Spenser and the Early Poetry of Milton*, PMLA, Vol. 98, No. 1, 1983

Guillaume, G., *The Prologues of the Lay le Freine and Sir Orfeo*, Modern Language Notes, Vol. 36, No. 8, The Johns Hopkins University Press 1921

Hall, A., *Elves in Anglo–Saxon England – Matters of Belief, Health, Gender and Identity*, The Boydell Press 2007

Hall, E.; Uhr, H., *Aureola and Fructus: Distinctions of Beatitude in Scholastic Thought and the Meaning of Some Crowns in Early Flemish Painting*, The Art Bulletin, Vol. 60, No. 2, College Art Association 1978

Hamilton, A.M., *The Singing Silence*, Phares 2007

Hamilton, A.M., *The Winging Word*, Phares 2008

Hamilton, A.M., *The Listening Land*, Phares 2009

Hamilton, G.L., *"Capados," and the Date of "Sir Gawayne and the Green Knight"*, Modern Philology, Vol. 5, No. 3, The University of Chicago Press 1908

Hanford, J.H.; *Dame Nature and Lady Life*, Modern Philology, Vol. 15, No. 5, The University of Chicago Press 1917

Hanford, J.H.; Steadman, J.M.(eds.), *Death and Liffe: An Alliterative Poem*, Chapel Hill 1918

Higgins, P.M., *Mathematics for the Curious*, Oxford University Press, 1998

Harrison, K. (tr.), *Sir Gawain and the Green Knight*, Oxford World's Classics, 1998

Harwood, B.J., *Gawain and the Gift*, PMLA, Vol. 106, No. 3., 1991

Hawking, S. (ed.), *God Created the Integers – The Mathematical Breakthroughs that Changed History*, Running Press Book Publishers 2005

Hearn, M. F., *Conferences Celebrating the Nine Hundredth Anniversary of the Beginning of Durham Cathedral: "Engineering a Cathedral," and "Anglo-Norman Durham, 1093-1993"* The Journal of the Society of Architectural Historians, Vol. 53, No. 4 Society of Architectural Historians, 1994

Heather, P. J., *Precious Stones in the Middle-English Verse of the Fourteenth Century*, I Folklore, Vol. 42, No. 3, Taylor & Francis, Ltd. 1931

Heather, P. J., *Precious Stones in the Middle-English Verse of the Fourteenth Century, II. (Continued)*, Folklore, Vol. 42, No. 4 Taylor & Francis, Ltd. 1931

Heather, P. J., *Colour Symbolism: Part I*, Folklore, Vol. 59, No. 4 Taylor & Francis, Ltd. 1948

Heather, P. J., *Colour Symbolism: Part II*, Folklore, Vol. 60, No. 1, Taylor & Francis, Ltd. 1949

Heather, P. J., *Colour Symbolism: Part III*, Folklore, Vol. 60, No. 2, Taylor & Francis, Ltd. 1949

Heather, P. J., *Colour Symbolism: Part IV*, Folklore, Vol. 60, No. 3, Taylor & Francis, Ltd. 1949

Heller, S., *Light as Glamour: The Luminescent Ideal of Beauty in the Roman de la Rose*, Speculum, Vol. 76, No. 4, Medieval Academy of America 2001

Helm, J., *Eric and Enide: Cosmic Measures in Nature and the Hebrew Heritage* in Robert J. Surles, *Medieval Numerology: A Book of Essays*, Garland Publishing, 1993

Helm, J., *A Trick That Has Been Lost: Modern Society And Poetry Through The Eyes Of Calliope And Pythagoras In The Poetry of A.D. Hope*, Quadrant, July 2001 http://www.articlearchives.com/humanities-social-science/literature-literature/247843-1.html

Heng, G., *Feminine Knots and the Other Sir Gawain and the Green Knight*, PMLA, Vol. 106, No. 3, Modern Language Association 1991

Heninger, Jr., S. K.; Hart, T.E.; Laferriere, D.; Peterson, R. G.. *Measure and Symmetry in Literature* Vol. 92, No. 1 Modern Language Association 1977

Herity, M., *Carpet Pages and Chi-Rhos: Some Depictions in Early Irish Christian Manuscripts and Stone Carvings*, http://www.celt.dias.ie/publications/celtica/c21/c21-208.pdf (accessed 12/08/09)

Herz–Fischler, R., *A "Very Pleasant Theorem"*, The College Mathematics Journal, Vol. 24, No. 4, Mathematical Association of America 1993

Herz–Fischler, R., *A Mathematical History of the Golden Number*, Dover, 1998

Hibbard, L. A., *The Books of Sir Simon de Burley, 1387*, Modern Language Notes, Vol. 30, No. 6, The Johns Hopkins University Press 1915

Hierocles, *The Golden Verses of Pythagoras*, Concord Grove Press 1983

Hildegard of Bingen; Hart, C., (tr.) Bishop, J. (tr.), *Scivias*, Paulist Press, 1990

Hill, J., *A Sequence of Associations in the Composition of Christ 275–347*, The Review of English Studies, New Series, Vol. 27, No. 107., 1976

Hill, L.L., *Madden's Divisions of Sir Gawain and the 'Large Initial Capitals' of Cotton Nero A.X.*, Speculum, Vol. 21, No. 1. Jan., 1946

Hills, D.F., *Gawain's Fault in Sir Gawain and the Green Knight [Critical Studies of Sir Gawain and the Green Knight*, ed. Donald R. Howard and Christian K. Zacker] University of Notre Dame Press 1970

Hilmo, M., *Medieval Images, Icons, and Illustrated English Literary Texts from the Ruthwell Cross to the Ellesmere Chaucer* Ashgate, 2004

Hindman, S., *Sealed in Parchment – Rereadings of Knighthood in the Illuminated Manuscripts of Chrétien de Troyes*, The University of Chicago Press 1994

Holman, C.H., *Marere$_3$ Mysse in The Pearl*, Modern Language Notes, Vol. 66, No. 1, Jan., 1951

Hopper, G.M., *The Ungenerated Seven as an Index to Pythagorean Number Theory*, The American Mathematical Monthly, Vol. 43, No. 7, Mathematical Association of America 1936

Horgan, A. D., *Justice in The Pearl*, The Review of English Studies, New Series, Vol. 32, No. 126, Oxford University Press 1982

Howard, D.R., *Structure and Symmetry in Sir Gawain [Critical Studies of Sir Gawain and the Green Knight*, ed. Donald R. Howard and Christian K. Zacker] University of Notre Dame Press 1970

Howlett, D., *Rubisca: An Edition, Translation and Commentary*, Peritia – Journal of the Medieval Academy of Ireland, Vol. 10, 1996

Howlett, D., *Seven Studies in Seventh Century Texts*, Peritia – Journal of the Medieval Academy of Ireland, Vol. 10, 1996

Howlett, D., *British Books in Biblical Style*, Four Courts Press, 1997

Howlett, D., *Medius as 'middle' and 'mean'*, Peritia – Journal of the Medieval Academy of Ireland, Vol. 13, 1999

Howlett, D., *Dicuill on the Islands of the North*, Peritia – Journal of the Medieval Academy of Ireland, Vol. 13, 1999

Howlett, D., *More Israelite Learning in Insular Latin*, Peritia – Journal of the Medieval Academy of Ireland, Vol. 13, 1999

Hulbert, J. R., *Syr Gawayn and the Grene Knyzt. (Continued)*, Modern Philology, Vol. 13, No. 8, The University of Chicago Press 1915

Hulbert, J. R., *Syr Gawayn and the Grene Knyzt (Concluded)*, Modern Philology, Vol. 13, No. 12, The University of Chicago Press 1916

Hulbert, J. R., *The "West Midland" of the Romances*, Modern Philology, Vol. 19, No. 1, The University of Chicago Press 1921

Hulbert, J. R., *A Hypothesis concerning the Alliterative Revival*, Modern Philology, Vol. 28, No. 4, The University of Chicago Press 1931

Hulbert, J. R., *Quatrains in Middle English Alliterative Poems*, Modern Philology, Vol. 48, No. 2, The University of Chicago Press 1950

Hunt, A., *Gawain and the Green Chapel*, 2005

Huntley, H.E., *The Divine Proportion*, Dover Publications 1970

Hyman, S.E., *The Ritual View of Myth and the Mythic*, The Journal of American Folklore, Vol. 68, No. 270, Myth: A Symposium, University of Illinois Press 1955

Jackson, K.H., *A Celtic Miscellany – Translations from the Celtic Literature*, Penguin Books 1971

Jefferson, L., *Tournaments, Heraldry and the Knights of the Round Table*, Arthurian Literature XIV, D.S. Brewer 1996

Jones, E.A., *'Loo, Lordes Myne, Heere Is a Fit!': The Structure of Chaucer's Sir Thopas*, The Review of English Studies, New Series, Vol. 51, No. 202, Oxford University Press 2000

Jones, E.J., *An Examination of the Authorship of the Deposition and Death of Richard II attributed to Creton*, Speculum, Vol. 15, No. 4, Medieval Academy of America 1940

Joost–Gaugier, C.L., 2006, *Measuring Heaven: Pythagoras and His Influence on Thought and Art in Antiquity and the Middle Ages*, Cornell University Press

Kalvesmaki, J., *Number Symbolism In The Mediterranean Before A.D. 1000: A Select Bibliography*, http://www.kalvesmaki.com/Arithmetic/index.htm (accessed 23/07/08)

Kalvesmaki, J., *Ancient Number Symbolism: Glossary*, http://www.kalvesmaki.com/Arithmetic/index.htm (accessed 23/07/08)

King, M., *Te Ao Hurihuri: Aspects of Maoritanga*, Reed Books 1975

Kiteley, J.F., *The Knight Who Cared For His Life* [*Critical Studies of Sir Gawain and the Green Knight*, ed. Donald R. Howard and Christian K. Zacker] University of Notre Dame Press 1970

Kittredge, G.L., *Disenchantment by Decapitation*, The Journal of American Folklore, Vol. 18, No. 68, University of Illinois Press 1905

Kottler, B.; Markman, A.M., *A Concordance to Five Middle English poems: Cleanness, St. Erkenwald, Sir Gawain and the Green Knight, Patience & Pearl*, University of Pittsburgh Press 1966

Knight, T.E., *The Use of Aletheia for the "Truth of Unreason": Plato, the Septuagint, and Philo*, The American Journal of Philology, Vol. 114, No. 4, The Johns Hopkins University Press 1993

Krueger, J.R., *A Note on Alexander's Arabic Epithet*, Journal of the American Oriental Society, Vol. 81, No. 4. (Sep. – Dec., 1961)

Kökeritz, H., *Sir Gawain and the Green Knight, 1954*, Modern Language Notes, Vol. 58, No. 5., 1943

Kraeling, E.G., *The Handwriting on the Wall*, Journal of Biblical Literature, Vol. 63, No. 1., 1944

Lankham, I., *Some Common Mathematical Symbols and Abbreviations (with History)* http://www.math.ucdavis.edu/~issy/contact_info.html

Lawlor, R., *Pythagorean Number as Form, Color and Light*, Lindisfarne Letter 14, 1982

Lawton, D.A., *The Unity of Middle English Alliterative Poetry*, Speculum, Vol. 58, No. 1., 1983

Lease, E. B., *The Number Three, Mysterious, Mystic, Magic*, Classical Philology, Vol. 14, No. 1, The University of Chicago Press 1919

Lefferts, P.M., Bent, M., Bowers, R., Everist, M., Wathey, A., *New Sources of English Thirteenth – and Fourteenth–Century Polyphony*, Early Music History, Vol. 2, Cambridge University Press 1982

Lewis, C.S., *The Allegory of Love: A Study in Medieval Tradition*, Oxford University Press, 1936

Lewis, C.S., *The Discarded Image – An Introduction to Medieval and Renaissance Literature*, Cambridge University Press, 1964

Lewis, C.S., *The Anthropological Approach* [*Critical Studies of Sir Gawain and the Green Knight*, ed. Donald R. Howard and Christian K. Zacker] University of Notre Dame Press 1970

Lewis, C.S., *Christianity and Literature in Religion* and *Modern Literature: Essays in Theory and Criticism*, G.B. Tennyson & Edward E. Ericson Jr (eds.), Eerdmans, 1975

Lewis, J.S., *Gawain and the Green Knight*, *College English*, Vol. 21, No. 1., Oct., 1959

Liberman, A., *Word Origins...and how we know them*, Oxford University Press 2005

Lindahl, C., McNamara, J., Lindow, J., *Medieval Folklore – A Guide to Myths, Legends, Tales, Beliefs and Customs*, Oxford University Press 2002

Livingstone, S., *Scottish Festivals*, Birlinn 1997

Livio, M., *The Golden Ratio*, Broadway 2003

Loomis, L.H., *Gawain and the Green Knight*, [*Critical Studies of Sir Gawain and the Green Knight*, ed. Donald R. Howard and Christian K. Zacker] University of Notre Dame Press 1970

Loomis, L.H., *The Auchinleck Manuscript and a Possible London Bookshop of 1330–1340*, PMLA, Vol. 57, No. 3, Modern Language Association 1942

Loomis, R.S., *Gawain, Gwri, and Cuchulinn*, PMLA, Vol. 43, No. 2, Modern Language Association 1928

Loomis, R.S., *Geoffrey of Monmouth and Arthurian Origins*, Speculum, Vol. 3, No. 1, Medieval Academy of America 1928

Loomis, R.S., *Calogrenanz and Crestien's Originality*, Modern Language Notes, Vol. 43, No. 4, The Johns Hopkins University Press, 1928

Loomis, R.S., *Some Names in Arthurian Romance*, PMLA, Vol. 45, No. 2., Modern Language Association 1930

Loomis, R.S., *The Visit to the Perilous Castle: A Study of the Arthurian Modifications of an Irish Theme*, PMLA, Vol. 48, No. 4, Modern Language Association 1933

Loomis, R.S., *Sir Orfeo and Walter Map's De Nugis*, Modern Language Notes, Vol. 51, No. 1, The Johns Hopkins University Press 1936

Loomis, R.S., *The Spoils of Annwn: An Early Arthurian Poem*, PMLA, Vol. 56, No. 4, Modern Language Association 1941

Loomis, R.S., *King Arthur and the Antipodes*, Modern Philology, Vol. 38, No. 3., 1941

Loomis, R.S., *From Segontium to Sinadon – the Legends of a Cite Gaste*, Speculum, Vol. 22, No. 4, Medieval Academy of America 1947

Loomis, R.S., *The Grail Story of Chrétien de Troyes as Ritual and Symbolism*, PMLA, Vol. 71, No. 4, Modern Language Association 1956

Loomis, R.S., *Celtic Myth and Arthurian Romance*, Constable 1993

Louis, K. R. R. G., *The Significance of Sir Orfeo's Self-Exile*, The Review of English Studies, New Series, Vol. 18, No. 71, Oxford University Press 1967

Lowes, J.L., *The Prologue to the Legend of Good Women as Related to the French Marguerite Poems, and the Filostrato*, PMLA, Vol. 19, No. 4, Modern Language Association, 1904

Lucas, E. C., *Daniel: Resolving the Enigma*, Vetus Testamentum, Vol. 50, Fasc. 1, BRILL 2000

Lucas, P.J., *Hautdesert in Sir Gawain and the Green Knight*, Neophilologus Vol 70, No. 2, 1986

Lund, N.W., *The Literary Structure of Paul's Hymn to Love,* Journal of Biblical Literature, Vol. 50, No. 4, The Society of Biblical Literature 1931

Luttrell, C., *Sir Gawain and the Green Knight and the Versions of Caradoc,* Forum For Modern Language Studies 1979

Luttrell, C., *The Folktale Element in Sir Gawain and the Green Knight,* Studies in Philology Vol LXXVII No 2 1980

Lynch, J. J., *The Prioress's Gems,* Modern Language Notes, Vol. 57, No. 6, The Johns Hopkins University Press 1942

Macaulay, A., *Apollo: The Pythagorean Definition of God,* Lindisfarne Letter 14, 1982

Macchioro, V., *Orphism and Paulinism,* The Journal of Religion, Vol. 8, No. 3, The University of Chicago Press 1928

Magness, J., *Heaven on Earth: Helios and the Zodiac Cycle in Ancient Palestinian Synagogues,* Dumbarton Oaks Papers, Vol. 59, Dumbarton Oaks 2005

Magoun, Jr., F. P., *Chaucer's Sir Gawain and the Ofr. Roman de la Rose,* Modern Language Notes, Vol. 67, No. 3, The Johns Hopkins University Press 1952

Mahan, M., *Palmoni, or the Numerals of Scripture,* D. Appleton and Company, 1863

Malarkey, S., Toelken, J.B., *Gawain and the Green Girdle* [*Critical Studies of Sir Gawain and the Green Knight,* ed. Donald R. Howard and Christian K. Zacker] University of Notre Dame Press 1970

Malone, K., *On the Etymology of Filch,* Modern Language Notes, Vol. 70, No. 3, The Johns Hopkins University Press 1955

Mandel, J., *The Ethical Context of Erec's Character,* The French Review, Vol. 50, No. 3, American Association of Teachers of French, 1977

Mann, J., *Price and Value in Sir Gawain and the Green Knight,* Essays in Criticism XXXVI: 294–318, 1986.

Manning, S., *A Psychological Interpretation of Sir Gawain and the Green Knight* [*Critical Studies of Sir Gawain and the Green Knight,* ed. Donald R. Howard and Christian K. Zacker] University of Notre Dame Press 1970

Markman, A.M., *The Meaning of Sir Gawain and the Green Knight,* PMLA, Vol. 72, No. 4., Sep., 1957

Markowsky, G., *Making a Golden Rectangle by Paper Folding,* The Mathematical Gazette, Vol. 75, No. 471, The Mathematical Association 1991

Markowsky, G., *Misconceptions about the Golden Ratio*, The College Mathematics Journal, Vol. 23, No. 1, 1992

Matarasso, P. M. (tr.), *Queste del Saint Graal,* Penguin Books, 1970

Matthews, C., *The Celtic Book of Days – A Celebration of Celtic Wisdom*, Godsfield Press Ltd, 1995

Matthews, J., *The Book of Arthur – Lost Tales from the Round Table*, Vega 2002

Matthews, J., *Gawain: The Knight of the Goddess – Restoring an Archetype* Aquarian Press 1990

Matthews, J.(ed.), *The Celtic Reader – Selections from Celtic Legend, Scholarship and Story* Aquarian/Thorsons 1991

Matthews, WH, *Mazes and Labyrinths: Their History and Development*, NABU Reprint of 1882 edition

Maw, D., *Machaut and the 'Critical' Phase of Medieval Polyphony* in *Music and Letters*, vol. 87, no. 2, Oxford University Press

McCarthy, C., *Sir Gawain and the Green Knight and the Sign of Trawpe,* Neophilologus, Vol. 85, No.2, 2001

McCarthy, C., *Luf–talkyng in Sir Gawain and the Green Knight*, Neophilologus Vol. 92, 2008

McCash, J.H.M., *Marie de Champagne and Eleanor of Aquitaine – A Relationship Reexamined*, Speculum, A Journal of Medieval Studies, (LIV, No.4), October 1979

McClain, J., *Observations on the Geometric Design of Saint–Yved at Braine*, Zeitschrift für Kunstgeschichte, 49 Bd., H. 1, Deutscher Kunstverlag GmbH Munchen Berlin 1986

McClure, P., *Gawain's Mesure and the Significance of the Three Hunts in Sir Gawain and the Green Knight*, Neophilologus Vol 57, No. 4, 1973

McGough, R.A., *The Bible Wheel: A Revelation of the Unity of the Divine Bible*, Bible Wheel Book House, 2006

Mead, W.E., *Color in Old English Poetry*, PMLA, Vol. 14, No. 2., 1899

Menken, M. J. J., *Numerical Literary Techniques in John – The Fourth Evangelist's Use of Numbers of Words and Syllables*, Supplements to Novum Testamentum, Vol LV, E.J. Brill, 1985

Menken, M. J. J., *The Position of σπλαγχνίζεσθαι and σπλάγχνα in the Gospel of Luke*, Novum Testamentum, Vol. 30, Fasc. 2, BRILL 1988

Menken, M. J. J., *The Textual Form of the Quotation from Isaiah 7:14 in Matthew 1:23*, Novum Testamentum, Vol. 43, Fasc. 2, BRILL 2001

Menninger, K., *Number Words and Number Symbols: A Cultural History of Numbers*, The M.I.T. Press, 1969.

Messier; Maty, M.(tr.), *A Memoir, Containing the History of the Return of the Famous Comet of 1682, with Observations of the Same, Made at Paris, at the Marine Observatory, in January, February, March, April, May, and the Beginning of June, 1759. By Mr. Messier, Astronomer, Keeper of the Journals, Plans, and Maps Belonging to the Marine of France, Fellow of the Royal Society in London, and Member of the Society of Sciences in Holland; Translated from the French by Matthew Maty, M.D. Sec. R. S.*, Philosophical Transactions (1683–1775), Vol. 55. 1765

Meyer, A.R., *Medieval Allegory and the Building of the New Jerusalem*, D.S. Brewer, 2003

Middleton, R. D., *Logos and Shekinah in the Fourth Gospel*, The Jewish Quarterly Review, New Series, Vol. 29, No. 2, University of Pennsylvania Press 1938

Miller, E. L., *The Logos of Heraclitus: Updating the Report*, The Harvard Theological Review, Vol. 74, No. 2, Cambridge University Press 1981

Miller, G. A., *A Wide-Spread Error Relating to the Pythagoreans*, Science, New Series, Vol. 82, No. 2119, American Association for the Advancement of Science 1935

Mills, M., *Christian Significance and Romance Tradition in Sir Gawain and the Green Knight* [*Critical Studies of Sir Gawain and the Green Knight*, ed. Donald R. Howard and Christian K. Zacker] University of Notre Dame Press 1970

Mills, M., (ed.), *Ywain and Gawain, Sir Percyvell of Gales, The Anturs of Arther*, Everyman 1992

Moffat, A., *Arthur and the Lost Kingdoms*, Orion Publishing 2000

Moorman, C., *The Role of the Narrator in "Patience"*, Modern Philology, Vol. 61, No. 2, The University of Chicago Press 1963

Morgan, G., *The Validity of Gawain's Confession in Sir Gawain and the Green Knight*, The Review of English Studies, Oxford University Press, 1985

Morgan, G., *Sir Gawain and the Green Knight and the Idea of Righteousness*, Irish Academic Press, 1991

Morgan, R.W., *The Trojan Era*, reprinted in Matthews, J. (ed), *The Celtic Reader – Selections from Celtic Legend, Scholarship and Story* Aquarian/Thorsons 1991

Moses, L., *Legends by the Numbers: The Symbolism of Numbers in the "Secret History of the Mongols"*, Asian Folklore Studies, Vol. 55, No. 1, Asian Folklore Studies, Nanzan University, 1996

Mustonen, S., *Extension of Golden Section To Multiple-Partite Division Of A Line Segment*, www.survo.fi/papers/nsection.pdf (accessed Dec 2007)

Nolan, B.; Farley-Hills, D., *The Authorship of Pearl: Two Notes*, The Review of English Studies, New Series, Vol. 22, No. 87, 1971

Nitze, W.A., *The Romance of Erec, Son of Lac*, Modern Philology, Vol. 11, No. 4, The University of Chicago Press, 1914

Nitze, W.A., *More on the Arthuriana of Nennius*, Modern Language Notes, Vol. 58, No. 1, The Johns Hopkins University Press 1943

Nitze, W.A., *The Character of Gauvain in the Romances of Chrétien de Troyes*, Modern Philology, Vol. 50, No. 4, The University of Chicago Press 1953

Ogden, G. S., *The Mathematics of Wisdom: Qoheleth IV 1–12*, Vetus Testamentum, Vol. 34, Fasc. 4, BRILL 1984

Olitzky, Kerry M.; Isaacs, Ronald H., *I Believe: The Thirteen Principles of Faith: A Confirmation Textbook*, KTAV Publishing, 2003

O'Reilly, M., *Clonmelsh and Mathematics* – Fr Ingram Memorial Lecture, Carlow, 24th November 2000

Oruch, J.B., *St. Valentine, Chaucer, and Spring in February*, Speculum, Vol. 56, No. 3., 1981

Outler, A. C., *The "Platonism" of Clement of Alexandria*, The Journal of Religion, Vol. 20, No. 3, The University of Chicago Press 1940

Ozdural, A., *Omar Khayyam, Mathematicians, and "Conversazioni" with Artisans*, The Journal of the Society of Architectural Historians, Vol. 54, No. 1, Society of Architectural Historians 1995

Pardee, D., *yph̬ "Witness" in Hebrew and Ugaritic*, Vetus Testamentum, Vol. 28, Fasc. 2,

BRILL 1978

Paris, W.A., *"Heroic Struggle": A Medieval and Modern Dilemma*, Journal of Religion and Health, Vol. 27, No. 2, 1988

Parunak, H. V. D., The *Literary Architecture of Ezekiel's Mar,ôT ,Elōhîm*, Journal of Biblical Literature, Vol. 99, No. 1, The Society of Biblical Literature 1980

Paschal, M., *The Structure of the Roman d'Enéas*, The French Review, Vol. 54, No. 1, American Association of Teachers of French, 1980

Patch, H. R., *Some Elements in Mediaeval Descriptions of the Otherworld*, PMLA, Vol. 33, No. 4, Modern Language Association 1918

Patch, H. R., *Precious Stones in the House of Fame*, Modern Language Notes, Vol. 50, No. 5, The Johns Hopkins University Press 1935

Patterson, D., *Hebrew Language and Jewish Thought*, Routledge 2005

Paul, S. M., *Amos 1:3–2:3: A Concatenous Literary Pattern*, Journal of Biblical Literature, Vol. 90, No. 4, The Society of Biblical Literature, 1971

Peck, R. A., *The Careful Hunter in the Parlement of the Thre Ages*, ELH, Vol. 39, No. 3, The Johns Hopkins University Press 1972

Perry, A. M., *The Framework of the Sermon on the Mount*, Journal of Biblical Literature, Vol. 54, No. 2, The Society of Biblical Literature 1935

Peters, J. P., *Some Uses of Numbers*, Journal of Biblical Literature, Vol. 38, No. 1/2, The Society of Biblical Literature 1919

Petersen, N. R., *The Composition of Mark 4:1–8:26*, The Harvard Theological Review, Vol. 73, No. 1/2, Dedicated to the Centennial of the Society of Biblical Literature, Cambridge University Press 1980

Peterson, C. J., *Pearl and St. Erkenwald: Some Evidence for Authorship*, The Review of English Studies, New Series, Vol. 25, No. 97, Oxford University Press 1974

Peterson, C. J., *The Pearl–Poet and John Massey of Cotton, Cheshire*, The Review of English Studies, New Series, Vol. 25, No. 99, Oxford University Press 1974

Peterson, C.; Wilson, E., *Hoccleve, the Old Hall Manuscript, Cotton Nero A.x, and the Pearl–Poet*, The Review of English Studies, New Series, Vol. 28, No. 109., Oxford University Press 1977

Peterson, R.G., *Critical Calculations: Measure and Symmetry in Literature*, PMLA, Vol. 91, No. 3, Modern Language Association, 1976

Philip, J. A., *Aristotle's Monograph on the Pythagoreans*, Transactions and Proceedings of the American Philological Association, Vol. 94, The Johns Hopkins University Press 1963

Pickering, D., *Dictionary of Superstitions*, Cassell 1995

Pickering, O.S., *Newly Discovered Secular Lyrics from Later Thirteenth-Century Cheshire*, The Review of English Studies, New Series, Vol. 43, No. 170, Oxford University Press 1992

Pickering, O.S. (ed.), *Individuality and Achievement in Middle English Poetry*, D.S. Brewer, 1997

Pickover, C.A., *A Passion for Mathematics – Numbers, Puzzles, Madness, Religion and the Quest for Reality*, John Wiley & Sons Inc. 2005

Plato; Waterfield, R.A.H. (tr.), *Theaetetus*, Penguin Classics, 1997

Pooley, R., *What does literature do?* in *The Discerning Reader – Christian Perspectives on Literature and Theory*, Barratt, D., Pooley, R., Ryken, L. (eds.), Baker Books, 1995

Prak, N.L., *Measurements of Amiens Cathedral*, The Journal of the Society of Architectural Historians, Vol. 25, No. 3, Society of Architectural Historians 1966

Puhvel, M., *Art and the Supernatural in Sir Gawain and the Green Knight*, ARTHURIAN LITERATURE V, D.S. Brewer 1985

Pulsiano, P., Wolf, K., *Medieval Scandinavia: An Encyclopedia*, Garland Publishing, 1993

Quasten, J., *A Pythagorean Idea in Jerome*, The American Journal of Philology, Vol. 63, No. 2, The Johns Hopkins University Press 1942

Quinn, W. A., *A Liturgical Detail and an Alternative Reading of St. Erkenwald, Line 319*, The Review of English Studies, New Series, Vol. 35, No. 139, Oxford University Press 1984

Raffel, B. (tr.), *Poems from the Old English* (Second Edition) University of Nebraska Press 1964

Raffel, B. (tr.), *Sir Gawain and the Green Knight*, Signet Classics, 1970

Raffel, B. (tr.), *Yvain: The Knight of the Lion*, Yale University Press 1987

Raffel, B. (tr.), *Erec and Enide*, Yale University Press 1997

Raffel, B. (tr.), *Lancelot: The Knight of the Cart*, Yale University Press 1997

Raffel, B. (tr.), *Cligés*, Yale University Press 1997

Raffel, B. (tr.), *Poems and Prose from the Old English*, Yale University Press 1998

Randall, D. B. J., *A Note on Structure in Sir Gawain and the Green Knight*, Modern Language Notes, Vol. 72, No. 3. Mar., 1957

Ramsay, L., *The French Book in Chivalric Romances: Popular legend in Medieval England*, Indiana University Press, Bloomington, 1983,

Reader, W. W., *The Twelve Jewels of Revelation 21:19–20: Tradition, History and Modern Interpretations*, Journal of Biblical Literature, Vol. 100, No. 3, The Society of Biblical Literature 1981

Reichardt, P.F., *Gawain and the Image of the Wound*, PMLA, Vol. 99, No. 2, Modern Language Association 1984

Reid, PV, *Moses's staff and Aeneas's shield: the way of the Torah versus classical heroism*, University Press of Arizona

Renoir, A., *An Echo to the Sense: The Patterns of Sound in Sir Gawain and the Green Knight* [*Critical Studies of Sir Gawain and the Green Knight*, ed. Donald R. Howard and Christian K. Zacker] University of Notre Dame Press 1970

Renoir, A., *A Minor Analogue of Sir Gawain and the Green Knight*, Neophilologus Vol 44, No. 1, 1960

Revard, C, *Was the Pearl poet in Aquitaine with Chaucer? A note on FADE, l.149 of Sir Gawain and the Green Knight*, SELIM 11

Rickard, M.R., *The Iconography of the Virgin Portal at Amiens*, Gesta, Vol. 22, No. 2, International Center of Medieval Art 1983

Robbins, F.E., *Posidonius and the Sources of Pythagorean Arithmology*, Classical Philology, Vol. 15, No. 4, The University of Chicago Press 1920

Robbins, F.E., *The Tradition of Greek Arithmology*, Classical Philology, Vol. 16, No. 2, The University of Chicago Press 1921

Robbins, F.E., *Arithmetic in Philo Judaeus*, Classical Philology, Vol. 26, No. 4, The University of Chicago Press 1931

Robertson, M., *Stanzaic Symmetry in Sir Gawain and the Green Knight*, *Speculum,*Vol. 57, No. 4, 1982

Runia, D.T., *Why Does Clement of Alexandria Call Philo "The Pythagorean"?* Vigiliae Christianae, Vol. 49, No. 1, BRILL 1995

Russell, H.G, *Lollard Opposition to Oaths by Creatures*, The American Historical Review, Vol. 51, No. 4, American Historical Association 1946

Russell, J.B, *A History of Medieval Christianity–Prophecy and Order*, Thomas Y. Crowell Company, 1968

Russell, J.B.; Alexander, B., *A New History of Witchcraft: Sorcerers, Heretics and Pagans*, Thames & Hudson, 2007

Sanderlin, G, *Two Transfigurations: Gawain and Aeneas*, The Chaucer Review, Vol 12, No 4, 1978

Savage, H.L., *A Note on Sir Gawain and the Green Knight 700–2*, Modern Language Notes, Vol. 46, No. 7. Nov., 1931

Savage, H.L., *Sir Gawain and the Order of the Garter*, ELH, Vol. 5, No. 2, The Johns Hopkins University Press 1938

Savage, H.L., *A Note on Sir Gawain 1795*, Modern Language Notes, Vol. 55, No. 8, The Johns Hopkins University Press 1940

Sayers, W., *Marie de France's Chievrefoil, Hazel Rods, and the Ogam Letters Coll and Uillenn,* Arthuriana Vol 14, No 2, 2004

Schleusener, J., *"Patience," Lines 35–40*, Modern Philology, Vol. 67, No. 1., 1969

Schleusener, J., *History and Action in Patience*, PMLA, Vol. 86, No. 5, Modern Language Association 1971

Schmidt, A. and Jacob, N. (eds.), *Introduction: Romance* in *Medieval English Romances*: Part One, Hodder and Stoughton, 1980

Schmidt, A. V. C., *'Latent Content' and 'The Testimony in the Text': Symbolic Meaning in Sir Gawain and the Green Knight*, The Review of English Studies, New Series, Vol. 38, No. 150, Oxford University Press 1987

Schoeck, R.J., *Mathematics and the Languages of Literary Criticism*, The Journal of Aesthetics and Art Criticism, Vol. 26, No. 3, Blackwell Publishing on behalf of The American Society for Aesthetics, 1968

Schofield, W.H., *The Nature and Fabric of The Pearl*, PMLA, Vol. 19, No. 1, Modern Language Association 1904

Schofield, W.H., *Symbolism, Allegory, and Autobiography in the Pearl*, PMLA, Vol. 24, No. 4, Modern Language Association 1909

Schrenk, L.P., *God as Monad: The Philosophical Basis of Medieval Theological Numerology* in Robert L. Surles (ed.), *Medieval Numerology: A Book of Essays*, Garland Publishing, 1993

Schweitzer, I., The Crux Gemmata and Shifting Significances of the Cross in Insular Art, http://www.marginalia.co.uk/journal/06illumination/schweitzer.php (accessed 04/09/08)

Scott, M.A., *Powlert: An Unexplained Folk-Song Word*, Modern Language Notes, Vol. 29, No. 4, The Johns Hopkins University Press 1914

Šedinova, H., *The Symbolism of the Precious Stones in St. Wenceslas Chapel*, Artibus et Historiae, Vol. 20, No. 39, IRSA s.c. 1999

Sibinga, J.S., *Ignatius and Matthew*, Novum Testamentum, Vol. 8, Fasc. 2/4, BRILL 1966

Sibinga, J.S., *Melito of Sardis. The Artist and His Text*, Vigiliae Christianae, Vol. 24, No. 2, BRILL 1970

Sibinga, J.S., *Some Observations on the Composition of Psalm XLVII*, Vetus Testamentum, Vol. 38, Fasc. 4., 1988

Sibinga, J.S., *Exploring The Composition Of Matth. 5–7: The Sermon On The Mount And Some Of Its "Structures"*, Filología Neotestamentaria 7, 1994

Sibinga, J.S., *Serta Paulina – On Composition Technique in Paul*, Filología Neotestamentaria 10, 1997

Sibinga, J.S., *1 Cor. 15:8/9 and Other Divisions in 1 Cor. 15:1–11*, Novum Testamentum, Vol. 39, Fasc. 1, BRILL 1997

Sibinga, J.S., *The Composition of 1 Cor. 9 and Its Context*, Novum Testamentum, Vol. 40, Fasc. 2, BRILL, 1998

Silverstein, T., *"Sir Gawain," Dear Brutus, and Britain's Fortunate Founding: A Study in Comedy and Convention*, Modern Philology, Vol. 62, No. 3, The University of Chicago Press 1965

Silverstein, T., *The Art of Sir Gawain and the Green Knight* [Critical Studies of Sir Gawain and the Green Knight, ed. Donald R. Howard and Christian K. Zacker] University of Notre Dame Press 1970

Silverstein, T., *Sir Gawain in a Dilemma, or Keeping Faith with Marcus Tullius Cicero*, Modern Philology, Vol. 75, No. 1, The University of Chicago Press 1977

Silverstein, T., *Sir Gawain and the Green Knight*, A New Critical Edition, University of Chicago Press 1974

Simpson, J., *Otherworld Adventures in an Icelandic Saga*, Folklore, Vol. 77, No. 1, Taylor & Francis, Ltd. 1966

Simson, O.G.v., *The Gothic Cathedral: Design and Meaning*, The Journal of the Society of Architectural Historians, vol. 11, no. 3, 1952

Sinclair, A., *The Discovery of the Grail*, Arrow Books, 1999

Slocum, K.B., *Speculum musicae: Jacques de Liège and the Art of Musical Number* in Robert L. Surles (ed.), *Medieval Numerology: A Book of Essays*, Garland Publishing 1993

Smit, J., *The Genre of 1 Corinthians 13 in the Light of Classical Rhetoric*, Novum Testamentum, Vol. 33, Fasc. 3, BRILL 1991

Smith, J.H., *Gawain's Leap: G. G. K. l. 2316*, Modern Language Notes, Vol. 49, No. 7. Nov., 1934

Solomon, J., *The Lesson of Sir Gawain* [Critical Studies of Sir Gawain and the Green Knight, ed. Donald R. Howard and Christian K. Zacker] University of Notre Dame Press 1970

Spallino, C., *Song of Solomon: An Adventure in Structure*, Callaloo, No. 25, Recent Essays from Europe: A Special Issue, The Johns Hopkins University Press 1985

Spearing, A.C., *Symbolic and Dramatic Development in "Pearl"*, Modern Philology, Vol. 60, No. 1, The University of Chicago Press 1962

Spearing, A.C., *Gawain's Speeches and the Poetry of "Cortaysye"* [Critical Studies of Sir Gawain and the Green Knight, ed. Donald R. Howard and Christian K. Zacker] University of Notre Dame Press 1970

Spearing, A.C., *Central and Displaced Sovereignty in Three Medieval Poems*, The Review of English Studies, New Series, Vol. 33, No. 131, Oxford University Press 1982

Spears, J.E., *The 'Boar's Head Carol' and Folk Tradition*, Folklore, Vol. 85, No. 3, Taylor & Francis, Ltd. 1974

Speirs, J., *Medieval English Poetry – The Non–Chaucerian Tradition*, Domville–Fife Press, 2007

Spencer, C. *The Heretic's Feast: A History of Vegetarianism*, UPNE, 1995

Stanbury, S., *The Body and the City in Pearl*, Representations, No. 48, University of California Press 1994

Stanbury, S., *Pearl*, Kalamazoo, Michigan: Medieval Institute Publications 2001

Stapleton, H.E.; G. J. W., *Ancient and Modern Aspects of Pythagoreanism*, Osiris, Vol. 13, The University of Chicago Press 1958

Stevenson, R.B.K., *Further notes on the Hunterston and Tara' brooches, Monymusk reliquary and Blackness bracelet*, Proceedings of the Society of Antiquities of Scotland 1983, http://archaeologydataservice.ac.uk/catalogue/adsdata/PSAS_2002/pdf/vol_113/113_469_477.pdf

Stevens, M., *Laughter and Game in Sir Gawain and the Green Knight*, Speculum, Vol. 47, No. 1, Medieval Academy of America 1972

Stevick, R. D., *The Length of Guthlac A*, Viator, Brepols Publishers, 1982

Stevick, R. D., *The 4×3 Crosses in the Lindisfarne and Lichfield Gospels*, Gesta, Vol. 25, No. 2 International Center of Medieval Art 1986

Stevick, R. D., *The Echternach Gospels' evangelist–symbol pages: forms from the 'two true measures of geometry'*, Peritia – Journal of the Medieval Academy of Ireland, Vol. 5, 1986

Stevick, R.D., *The Shapes of the Book of Durrow Evangelist–Symbol Pages*, The Art Bulletin, Vol. 68, No. 2, College Art Association 1986

Stevick, R.D., *The Harmonic Plan of the Harburg Gospels' Cross–Page*, Artibus et Historiae, Vol. 12, No. 23. 1991

Stevick, R.D., *The Earliest Irish and English Bookarts – Visual and Poetic Forms Before A.D. 1000*, University of Pennsylvania Press 1994

Stevick, R.D., *Shapes of Early Sculptured Crosses of Ireland*, Gesta, Vol. 38, No. 1., 1999

Stevick, R.D., *The Shape of the Durrow Cross*, Peritia – Journal of the Medieval Academy of Ireland, Vol. 13, 1999

Stevick, R. D., *The Coherent Geometry of Two Irish High Crosses*, Peritia – Journal of the Medieval Academy of Ireland, Vol. 14, 2000

Stewart, I., *Letters to a Young Mathematician*, Basic Books 2006

Stocks, J.L., *Plato and the Tripartite Soul*, Mind, New Series, Vol. 24, No. 94, Oxford University Press 1915

Stone, B. (tr.), *Sir Gawain and the Green Knight*, Penguin Books 1959

Stone, B. (tr.), *Medieval English Verse*, Penguin Books 1964

Stone, B. (tr.), *The Owl and the Nightingale/Cleanness/St. Erkenwald* Penguin Books 1988

Stroud, M., *Chivalric Terminology in Late Medieval Literature*, Journal of the History of Ideas, Vol. 37, No. 2, University of Pennsylvania Press 1976

Sumney, J.L., *The Letter of Eugnostos and the Origins of Gnosticism*, Novum Testamentum, Vol. 31, Fasc. 2, BRILL 1989

Surles, R.L., *Medieval Numerology: A Book of Essays*, Garland Publishing 1993

Swanson, R. N., *Universities, Graduates and Benefices in Later Medieval England*, Past and Present, No. 106, Oxford University Press 1985

Takagi, M., *Image Resurrection of Sir Gawain and the Green Knight in The Silver Chair: C.S. Lewis as a Medievalist* http://www.flet.keio.ac.jp/~colloq/articles/backnumb/Col_23_takagi.pdf (April 2008)

Tamplin, R., *The Saints in Sir Gawain and the Green Knight*, Medieval Academy of America 1969

Tamplin, R., *The Tempest and The Waste Land*, American Literature, Vol. 39, No. 3, Duke University Press 1967

Tavernor, R., 2007, *Smoot's Ear: The Measure of Humanity*, Yale University Press

Teicher, J. L., *The Christian Interpretation of the Sign × in the Isaiah Scroll*, Vetus Testamentum, Vol. 5, Fasc. 2, BRILL 1955

Terrien, S., *The Omphalos Myth and Hebrew Religion*, Vetus Testamentum, Vol. 20, Fasc.

3, BRILL, 1970

Tester, S.K., *The Use of the Word Lee in Sir Gawain and the Green Knight*, Neophilologus, Vol. 54, No. 1, 1970

Thacker, A., *The Cult of King Harold at Chester* (in *The Middle Ages in the North–West* ed. Tom Scott and Pat Starkey) Leopard's Head Press 1995

Thom, J.C., *The Journey Up and Down: Pythagoras in Two Greek Apologists*, Church History, vol.58, no. 3, September 1989

Thom, J.C., *"Don't Walk on the Highways": The Pythagorean Akousmata and Early Christian Literature,* Journal of Biblical Literature, 1994

Thom, J.C., *Dyads, Triads and Other Compositional Beasts in the Sermon on the Mount (Matthew 5–7)*, Supplements to Novum Testamentum, Vol. 124, ed. Breytenbach, C., Thom, J.C., Punt, J., *The New Testament Interpreted: Essays in Honour of Bernard C. Lategan,* BRILL 2006

Thomas, C., *The Llanddewi-Brefi 'Idnert' Stone*, Pertitia – Journal of the Medieval Academy of Ireland, Vol, 10, 1996

Thomas, C., *Whispering Reeds, or the Anglesey Catamanus Insription Stript Bare: A Detective Story*, Oxbow Books 2002

Thompson, R. H., *The Perils of Good Advice: The Effect of the Wise Counsellor upon the Conduct of Gawain*, Folklore, Vol. 90, No. 1, Taylor & Francis, Ltd. 1979

Toker, F., *Gothic Architecture by Remote Control: An Illustrated Building Contract of 1340*, The Art Bulletin, Vol. 67, No. 1, College Art Association 1985

Tolkien, J.R.R., *Some Contributions to Middle–English Lexicography*, The Review of English Studies, Vol. 1, No. 2, Oxford University Press 1925

Tolkien, J.R.R., *The Devil's Coach–Horses: Eaueres*, The Review of English Studies, Vol. 1, No. 3, Oxford University Press 1925

Tolkien, J.R.R., Gordon, E.V., *Sir Gawain and the Green Knight*, Clarendon Press 1963

Tolkien, J.R.R. (tr.), Tolkien, C. (ed.), 1975, *Sir Gawain and the Green Knight, Pearl and Sir Orfeo*, Unwin Paperbacks

Tolkien, J.R.R., *Sir Gawain and the Green Knight* in *The Monsters and the Critics and Other Essays*, Harper Collins 1983

Trapp, J.B.(ed.), 1973, *The Oxford Anthology of English Literature: Medieval English Literature*, Oxford University Press

Traver, H., *The Four Daughters of God, A Study of the Versions of this Allegory with Especial Reference to those in Latin, French and English*, The John C Winston Co, 1907

Travis, P.W., *Chaucer's Heliotropes and the Poetics of Metaphor*, Speculum, Vol. 72, No. 2, Medieval Academy of America 1997

Travis, W., *Representing "Christ as Giant" in Early Medieval Art*, Zeitschrift für Kunstgeschichte, 62 Bd., H. 2, Deutscher Kunstverlag GmbH Munchen Berlin 1999

Trumball, H.C., *The Salt Covenant*, Impact Christian Books, 1999

Turville-Petre, T., *'Summer Sunday', 'De Tribus Regibus Mortuis', and 'The Awntyrs off Arthure': Three Poems in the Thirteen-Line Stanza*, The Review of English Studies, New Series, Vol. 25, No. 97, Oxford University Press 1974

Turville-Petre, T., Wilson, E., *Hoccleve, 'Maistir Massy' and the Pearl Poet: Two Notes*, The Review of English Studies, New Series, Vol. 26, No. 102, Oxford University Press 1975

Tuttleton, J. W., *The Manuscript Divisions of Sir Gawain and the Green Knight*, Speculum, Vol. 41, No. 2., Apr., 1966

Verschuur, G.L., *Impact! The Threat of Comets and Asteroids*, Oxford University Press 1996

Walker, D. P., *Orpheus the Theologian and Renaissance Platonists*, Journal of the Warburg and Courtauld Institutes, Vol. 16, No. 1 / 2, The Warburg Institute 1953

Walker, D. P., *Kepler's Celestial Music*, Journal of the Warburg and Courtauld Institutes, Vol. 30, The Warburg Institute 1967

Weiss, M., *The Pattern of Numerical Sequence in Amos 1-2: A Re-Examination*, Journal of Biblical Literature, Vol. 86, No. 4, The Society of Biblical Literature 1967

Wenzel, S., *Reflections on (New) Philology*, Speculum, Vol. 65, No. 1, Medieval Academy of America 1990

Werner, M., *The Cross-Carpet Page in the Book of Durrow: The Cult of the True Cross,*

Adomnan, and Iona, The Art Bulletin, Vol. 72, No. 2, College Art Association 1990

Wertheim, M., *Pythagoras' Trousers*, W. W. Norton & Company 1997

Weston, J.L., *From Ritual to Romance: An account of the Holy Grail from ancient ritual to Christian Symbol*, Doubleday Anchor 1957

Weston, J.L., *The Quest of the Holy Grail*, Frank Cass & Co 1964

Weston, J.L., *The Three Days Tournament*, Haskell House 1965

Weston, J.L., *Sir Gawain and the Green Knight*, Dover Publications 2003

Whatley, G., *Heathens and Saints: St. Erkenwald in Its Legendary Context*, Speculum, Vol. 61, No. 2, Medieval Academy of America 1986

White, H., *Blood in Pearl*, The Review of English Studies, New Series, Vol. 38, No. 149. 1987

White Jr., R.B., *A Note on the Green Knight's Red Eyes* [*Critical Studies of Sir Gawain and the Green Knight*, ed. Donald R. Howard and Christian K. Zacker] University of Notre Dame Press 1970

Whittaker, J., *Ammonius on the Delphic E*, The Classical Quarterly, New Series, Vol. 19, No. 1, Cambridge University Press 1969

Williams, C., Lewis, C.S., *Taliessin through Logres/ The Region of the Summer Stars/ Arthurian Torso* W.B. Eerdmans Publishing Company 1974

Williams, D., *The Point of "Patience"*, Modern Philology, Vol. 68, No. 2, The University of Chicago Press 1970

Williams, J.G., *Number Symbolism and Joseph as Symbol of Completion*, Journal of Biblical Literature, Vol. 98, No. 1, The Society of Biblical Literature 1979

Winny, J., (ed. & tr.), *Sir Gawain and the Green Knight*, Broadview Literary Texts 2002

Witt, R.E., *Plotinus and Posidonius*, The Classical Quarterly, Vol. 24, No. 3/4, Cambridge University Press 1930

Witt, R.E., *The Plotinian Logos and Its Stoic Basis*, The Classical Quarterly, Vol. 25, No. 2, Cambridge University Press 1931

Wright, A.E., *Gold and Grace in Hartmann's Gregorius* in *Medieval Numerology – A Book of Essays*, Robert L. Surles (ed.), Garland Publishing 1993

Wytzes, J., *The Twofold Way II Platonic Influences in the Work of Clement of Alexandria*, Vigiliae Christianae, Vol. 14, No. 3, BRILL 1960

Yoffie, L. R. C., *Songs of the "Twelve Numbers" and the Hebrew Chant of "Echod mi Yodea"*, The Journal of American Folklore, Vol. 62, No. 246, University of Illinois Press 1949

Youngblood, R, *Divine Names in the Book of Psalms: Literary Patterns and Number Structures*, JANES 19, 1989

Zajonc, A.G., *The Two Lights*, Lindisfarne Letter 14, 1982

Zöllner, F., *Agrippa, Leonardo and the Codex Huygens*, Journal of the Warburg and Courtauld Institutes, Vol. 48, The Warburg Institute 1985

Cambridge History of English and American Literature in 18 Volumes (1907–21), *Pearl, Cleanness, Patience, Sir Gawayne*, Volume I. From the Beginnings to the Cycles of Romance.

Index

alphanumeric — 2, 37, 59, 183, 185

Anderson, JJ — 19, 63, 68, 84, 91, 95, 164, 264, 278, 284, 289, 290, 309

Augustine of Hippo — 1, 2, 5, 6, 30, 31, 149, 162, 165, 261, 270, 320

Albertus Magnus — 4

Al–Biruni — 108

aliquot factor — 11, 12, 54, 123, 149, 165, 179, 261, 277

Al Khidr — 335

alliteration — 15, 40, 41, 46, 47, 54, 127, 129, 137, 154, 159, 166, 273, 299, 300

Anglo–Saxon Chronicle — 176

Anne of Bohemia — 42, 43, 53, 310

antiphraxis — 145

Apollo — 79, 138, 139, 144, 146, 154, 176, 177, 309, 316, 317, 320, 322, 340

Archimedes — 37, 44, 148, 202, 203, 274, 295, 319

architectonics — 261

arc length — 11, 13, 275, 295

arithmetic metaphor (see structural arithmetic metaphor)

arithmology/arithmetic theology — 195, 261

Arming Sequence — 17, 91, 132, 136, 160, 161, 168, 181, 200, 306

Armour of God — 23, 182–185, 306, 336, 337

Arthur, Ross — 121, 278, 284, 300, 306

Arundel, Thomas — 170, 186, 338

Asgrimsson, Eysteinn — 55

Ashton–on–Mersey — 20, 169, 199, 265, 266, 302

Atkinson, Keith — 7

Babylon — 16, 94, 154, 179, 299, 327, 328

Baillie, Mike — 175–177, 332, 334

Barraclough, Jane — 97

Barr, Mark — 13, 104, 112

Barron, WRJ — 140

Barrett, RW — 45, 336

Bartelt, Andrew — 261, 271

Bates, Linda — 173, 331

Hulme, Keri — 334

Hunterston Brooch — 112

Iamblichus — 144, 319

incarnation — 55, 57, 71, 74, 206

integer — 11, 12, 296

Isaiah — 3, 36, 261, 271, 276, 280, 331

Isocrates — 36, 111

Jean de Meun — 84, 85, 91, 140, 180

John of Salisbury — 115

St. John's Church, Chester — 128, 302

Jonah — 17, 48, 61, 63, 65, 150, 156, 158, 160, 162, 178, 289, 290

Jordan de Holme — 153, 199, 265, 302

St. Julian — 307

Julian bower (labyrinth) — 307

Julian calendar — 278

Julian of Norwich — 70, 93

Julian the Apostate — 143, 144, 319

Kamil, Abu — 108

Kells, Book of — 84, 112

Kemp Owayne — 310, 311

Kepler, Johannes — 106, 112, 295

Khayyam, Omar — 4

King, Michael — 334

Kosmos — 39, 277, 324

Lady Fortune or Dame Fortune — see Fortune

Lancashire — 186

Lancelot — 140, 160, 161, 175, 284, 314, 323

Lawrence, DH — 196, 261

Le Chevalier de la Charette — 86, 115, 175, 314

Left–hand numbers — 30, 77, 299

Leonardo da Vinci — 9, 26, 89, 111, 112, 268, 276, 287, 293

Levi, Éliphas — 111, 296

Lewis, CS — 4, 31, 262, 263, 270, 274, 325

Lilja — 55, 56, 58, 74, 137, 277

Lindisfarne gospels — 15, 24, 112, 116, 177

Logos — 37, 76, 79, 114, 144, 147

Lollards — 45, 199, 274, 338

Lot's wife — 180

Lucifer — 175, 331

Mabinogion, The — 150, 310

manicule — 155, 322

Marie de Champagne — 115

marguerites — 139, 140, 266

Mary Gipsy — 335

Masci, Masse, Massi, Massey, Macy, Hugh de/John — 19, 20, 75, 140, 169, 199, 264–266, 288, 302

McGough, Richard — 276

Menken, MJJ — 36, 114, 195, 261, 262, 267, 271, 283, 297, 318, 341

Merlin — 271, 335

Michaelmas — 66

Milton, John — 39, 167, 194, 291, 314, 316, 330, 339, 340

Moffat, Alistair — 336

Morgan the goddess —34, 35, 271, 335, 336

Muiredach Cross — 112, 116

Music of the Spheres — 32, 60, 84, 130, 131, 144, 191, 195, 201–293, 206, 262, 299, 317, 320, 341

Neo–Platonism — 5, 6, 38, 77, 83, 89, 93, 115–118, 120, 121, 137, 143, 144, 147, 159, 161, 162, 178, 187, 194, 198, 199, 204, 261, 287, 308, 314, 317, 319, 320, 324

Neo–Pythagoreanism — see Pythagoras

Nero — 15, 139, 309

New Jerusalem — 83, 90–92, 120, 153, 179, 202, 203, 296, 314

nirvana — 70, 71

numerator — 12, 13

numerical literary design/style/from/technique/structure — 3, 19, 59, 61, 90, 116, 180, 186, 114, 126, 149, 152, 166, 191, 194, 196, 199, 261, 271, 280, 282, 283, 297, 314, 318, 320, 340, 341

numerology — 3–5, 7, 19–21, 33, 37, 43, 59, 261, 262, 264, 267, 270, 272, 283, 287, 320, 326, 328

Odin — 335, 336

Ohm, Martin — 12, 106, 107, 187

Olsen, Jonathan — 339

Osiris — 145, 146, 339, 340

Owain Glendower — 185, 313

Owein the champion — see *Kemp Owayne*

Pacioli, Luca — 111, 268, 293

Paradise Lost — 195, 291, 339, 341

Parliament of Fowls — 19, 116, 118, 119, 122, 126, 147, 167, 188, 263, 272, 299, 300, 315, 330

Parthenon — 104

Patience — 15, 16, 30, 40, 43–46, 48, 49, 51, 53, 55–58, 60–74, 77, 82, 84, 85, 89, 92–96, 98, 100, 125, 127, 150, 156–160, 162, 164, 178, 179, 181, 198, 200, 208, 264, 269, 274, 275, 289, 290, 305, 307, 308, 313, 321, 323–325, 328–330, 336, 338

Paul of Tarsus — 23, 36, 76, 116, 131, 142, 146, 152–154, 182–185, 198, 317, 321, 322, 336

St. Paul's church — 126, 128, 305

Peard, Aya — 132

Pearl — 7–10, 15, 16, 18–20, 22, 24, 26, 29–33, 37–42, 44–55, 59–64, 68, 71–85, 88–92, 96, 97, 99, 103, 110, 112, 114, 115, 117, 119, 121, 122, 124–133, 136, 139–143, 145, 147–150, 152–161, 163, 165–170, 175, 178–180, 183, 185–187, 190–198, 200, 265–270, 272–275, 277, 279, 281–284, 286, 290, 292, 298–305, 308–310, 314–317, 320, 323–325, 328–331, 336, 338, 341

pentagon — 100, 101, 268, 294

pentagram — 24–28, 79, 101, 109–111, 117, 121, 123, 126, 127, 137, 138 163, 268, 296, 297, 299, 306, 327, 335

perfect numbers — 3, 12, 21, 35, 36, 43, 62, 64, 127, 162, 261, 277, 279

perimeter — 13, 56, 88, 89, 101, 102, 277, 283, 286, 287

Peterson, Clifford — 19, 75, 265

Peterson, RG — 4, 118, 298

Pheidias (Phidias) — 13, 104

Piers de Massy — 186

317, 318

statistics — 40, 59, 104

Stevick, Robert — 7, 112–114, 116, 262, 297

Stichometry — 35, 267, 315

Structural arithmetic metaphor — 3, 8, 26, 43, 51, 54, 62, 105, 125, 152, 188, 194, 261, 272, 278, 298, 304, 326

Summer Sunday — 4, 42, 43, 46–48, 52, 53, 121, 128, 129, 188, 198, 200, 275

Tau — 12–14, 107, 121, 129, 132, 134, 155–162, 181, 187, 201, 202, 205, 207, 300, 301, 315, 324, 328, 329, 337, 338

Tavenor, Robert — 297

Te Ao Hurihuri — 334

Te Kaihau — 334

Tetrakys (Tetract) — 14, 75, 79, 80, 133, 144, 145, 199, 203, 301, 320

Thierry of Chartres — 115

Timaeus — 3

Tolkien, JRR — 2, 97, 137, 190, 197, 263, 286, 292, 301, 305, 325

Torah — 35, 101, 306

The Touchstone — 336

Tournament — 85, 86, 186, 284, 285

tramountayne — 175

translatio studii — 141, 142

trawthe — 17, 97, 98, 109, 120, 121–123, 125, 126, 131, 133, 137, 148, 152, 156, 158, 159, 161, 162, 165, 168, 169, 178, 179, 182, 187, 194, 203, 204, 289, 292, 304, 305, 307, 316

tree rings — 175, 176

triangle, circumscribed — 56, 57, 74, 100, 101, 206, 277

triangular numbers — 14, 35–37, 54–56, 75, 76, 123, 126, 133, 149, 150, 158, 167, 181, 201–203, 261, 279, 301, 309, 320, 324, 326, 328, 336

triple triangle — 117, 123, 133, 148, 167, 179, 203, 296

Troy — 45, 141, 181, 263, 307, 308, 321

trwe as ston — 157, 158

Tschichold, Jan — 116

Turville-Petre — 265

endnotes

1   Augustine was not the only one to have such a counter–intuitive thought. Nicholas of Cusa also contemplated the notion of a relativistic universe.

2   So–called 'perfect' numbers are those numbers with aliquot factors (the number itself excluded) which add up to the number. The first three are 6, 28 and 496. The factors of 6 are 1, 2, 3 and 6. 1 + 2 + 3 = 6. The factors of 28 are, 1, 2, 4, 7, 14 and 28. 1 + 2 + 4 + 7 + 14 = 28.

3   Robert Tavernor, *Smoot's Ear: The Measure of Humanity*, Yale University Press, 2007

4   Andrew Bartelt in *The Book around Immanuel: Style and Structure in Isaiah 2–12* points out that there is a chiastic structure enclosing the prophecy about Immanuel. This takes the form of two parallel poems each of 496 syllables. 496 is the third 'perfect' number. This style of enclosure was apparently imitated by John, the writer of the fourth gospel. The Greek text of his gospel has 496 syllables in the prologue and 496 words in the concluding section.

5   A term used by MMJ Menken and J. Smit Sibinga. Other terms for similar combinations of words and numbers are 'Biblical style' (Howlett), 'structural arithmetic metaphor' (Bulatkin), 'architectonics' (Crawford), 'arithmology' or 'arithmetic theology' (Røstvig). I will use most of these and occasionally also the term 'mathematical token' since, in my opinion, various medieval romances use particular numbers like the tokens of favour bestowed in the romances of courtly love. I will also on occasion use the phrase 'mathematical grammar' to describe the totality of an author's work in one particular poem. Nevertheless, the terms are basically interchangeable. None of them, however, refer to numerology proper.

6   Edmund Spenser's use of it has been extraordinarily well–documented from the time of A. Kent Hieatt's analysis of his *Epithalamion* in 1960. In more recent times, however, poets have been far more sparing of their use of it. The only possible examples by well–known poets of which I am aware are: *The Chambered Nautilus* by Oliver Wendell Holmes (which, like the broken shell which inspired it, appears to have a fractured Fibonacci–like pattern) or *Tortoise Shell* by DH Lawrence (which not only advertises its neo–Platonic inspiration in the irregularity of its verses but also in its reference to Pythagoras). Indeed *Tortoise Shell* is built using the triangular numbers and the golden section.

7   WF Bolton (ed), *The Middle Ages – Volume 1 of the Penguin History of Literature* Penguin 1993 *p. 17*

8   CS Lewis, *The Discarded Image – An Introduction to Medieval and Renaissance Literature*, Cambridge University Press, 1964, p. 163

9   RJ Schoeck, *Mathematics and the Languages of Literary Criticism*, The Journal of Aesthetics and Art Criticism, Vol. 26, No. 3, Blackwell Publishing on behalf of The American Society for Aesthetics, 1968

10   RR Edwards, *Ratio and Invention, A Study of Medieval Lyric and Narrative*, Vanderbilt University Press, 1989, p.140 Significantly, Edwards maintains it is meant as a natural wonder, rather than a magical device.

11   Paul Aker in *The Emergence of an Arithmetical Mentality in Middle English Literature* (The Chaucer Review, Vol 28, No 3, 1994), commenting on Derek Brewer's *Arithmetic and the Mentality of Chaucer*.

12   This kind of 'fix' reminds me of Arthur Koestler's study of the medieval Polish text of Nicholas Copernicus' famous *De revolutionibus orbium coelestium—On the Revolution of the Heavenly Spheres*. Koestler was familiar with medieval Polish and, in his work *The Sleepwalkers: A History of Man's Changing Vision of the Universe*, maintained that, far from being a forward–thinking revolutionary, Copernicus was actually a reactionary. Moreover, Koestler claims the ubiquitous and famous diagram of the sun–centred solar system is a 'fix'—Copernicus' own diagram has the earth and not the sun at the centre! Subsequent editors have made the change to ensure that there is no discrepancy between what history tells us Copernicus wrote and what he actually did! Despite heavy criticism of Koestler as a popular writer whose research was biased, scholars still admit that he made a case which is neither easily answered, nor easily dismissed.

13   Jeffrey Burton Russell, *Exposing Myths About Christianity: A Guide to Answering 145 Viral Lies and Legends*, IVP Books 2012

14   Marcus du Santoy, *The Music of the Primes: Why an Unsolved Problem in Mathematics Matters*, Fourth Estate 2003

15   For the most part, the work of scholars such as Joan Helm, Ed Condren, Robert Stevick, David Howlett, Eleanor Bulatkin, George Duckworth, Thomas Elwood Hart, Thomas Hill, Charles Thomas, A. Kent Hieatt, Maarten Menken and J. Smit Sibinga which covers this wide spectrum from the poets of classical antiquity through the gospel writers to the early Renaissance tends to suggest that the

mathematics is there for reasons of aesthetics, self–authentication of the text or numerology. Although these scholars have made careful and thoughtful cases, their critics seem to imply that 'maths for math's sake' is what comes across. In a number of instances I tend to agree that the scholars have expected the mathematics to speak for itself. It's as if they have anticipated readers will automatically share their astonishment. However, without an appreciation of mathematics, the design is almost opaque. Vishal Mangalwadi suggests, in twenty–first century culture, that despite our protests to the contrary, there is no such thing as worldview. Our culture is dominated by a 'silo' mentality: mathematics in one silo, literature in another. It's impossible, he contends, to have a worldview inside a silo, because by its very nature, a silo excludes the world. In attempting to explore a worldview of the past, I have tried, not always successfully, to move beyond my own silo in order to find the inspiration behind a mathematical design.

16 And, for that matter, the first decade of the new millennium according to some opinion polls. No, it wasn't not JK Rowling (who surprisingly only gets on the list once with *Harry Potter and the Half–Blood Prince* at #5.) The authors were JRR Tolkien and CS Lewis.

17 Roger Pooley, *What does literature do?* in *The Discerning Reader – Christian Perspectives on Literature and Theory*, David Barratt, Roger Pooley, Leland Ryken (eds.), Baker Books, 1995

18 Kenneth Gross, *'Each Heav'nly Close': Mythologies and Metrics in Spenser and the Early Poetry of Milton*, PMLA, Vol. 98, No. 1, 1983

19 Author of the dream vision, *Piers Plowman*.

20 Author of *Speculum Meditantis*, *Vox Clamantis* and *Confessio Amantis*.

21 Author of the *Troy Book*, *Siege of Thebes* and *Fall of Princes*, John Lydgate of Bury St Edmunds was an admirer of Geoffrey Chaucer and friend of his son, Thomas Chaucer.

22 Most famous for *The Canterbury Tales* but also the author of other notable works, including *The Book of the Duchess*, *Parliament of Fowls*, *Troilus and Criseyde* and *The Legend of Good Women*.

23 Theodore Silverstein, *Sir Gawain and the Green Knight*, A New Critical Edition, University of Chicago Press 1974, p.111

24 To divide comedy from theology is appropriate in the twenty–first century with

its silos of knowledge (as Vishal Mangalwadi calls them). It is not appropriate to the fourteenth century. Despite the motivation of the killer in Umberto Eco's medieval crime thriller *The Name of the Rose* (a motivation in which the desire to deny the world access to humour and, in particular, Aristotle's lost treatise on comedy is paramount), it is really a post–modern take on the Middle Ages, projecting onto it a high–minded seriousness that may well be little more than a caricature.

25 Several studies have recognised the importance of the Four Daughters of God within confined sections of the manuscript. None have however, as far as I am aware, suggested it is the overall theme of the four poems. Nor have any analysed the mathematics to show the critical positioning of the four virtues within the text. JJ Anderson, for instance, sees the Four Daughters of God as integral to the two poems showing the most fundamental connection for him: *Cleanness* and *Patience*. 'The two poems contrast each other and may be seen as reciprocal' and 'In terms of this debate, *Cleanness* represents God as justice and truth, *Patience* as peace and mercy.' (*Language and Imagination in the Gawain–Poems*, Manchester University Press, 2005, p5) Although truth is the natural ally of justice (or righteousness) and peace the natural ally of mercy, Anderson's elision of them is, I think, premature. He also maintains (p14) that the poet nowhere actually alludes to the four daughters, although Andrew and Waldron clearly suggest otherwise.

26 I have only picked up some of his more subtle mathematics as I have tried to emulate his methodology, rather than his actual numbers, in my own fiction. Certainly Chaucer would be a candidate for a jury of peers (as shown in Ed Condren's analysis of *Troilus and Criseyde* where the positioning of 'dulcarnoun'—a reference to the famous Pythagorean theorem of right–angled triangles—is too fitting to be coincidental: *Chaucer from Prentice to Poet: The Metaphor of Love in Dream Visions and 'Troilus and Criseyde'*, Gainesville, Fla: University Press of Florida, 2008) There are curious echoes in the mathematical design of some of Chaucer's poems, suggesting either that they were using variations on a 'formula' or a kind of poetic battle.

27 *Language and Imagination in the Gawain–Poems*, Manchester University Press, 2005, p4

28 Moreover, John de Masci adds up to 101 using standard numerological replacements of letters and numbers.

29 John M. Bowers, *The politics of Pearl: court poetry in the age of Richard II*, p10

30 In Derek Brewer & Jonathan Gibson, *A companion to the Gawain-poet*, p 30

31 Lines 901–912. This has long been regarded by many scholars as an anomalous stanza in its own right.

32 Thomas Hoccleve, a contemporary of the *Pearl* poet, in a poem witten about 1411–1414 mentions a 'maister Massy', a man of 'fructuous intelligence'. Peterson, Wilson and Farley–Hills suggested that this is a reference to the 'intelligent and inaccessibly arcane' nature of John Massy's poetry, as does Roy Neil Graves who indicates the possibility of a 21–line runic poem embedded within *Pearl* which may encode the name Massy, Massey or Masci. Turville–Petre on the other hand considered that Hoccleve was simply addressing a potentially generous benefactor by the name of William Massy.

While I'm not sure that either Graves or Peterson has proved the case beyond reasonable doubt, I regard the points they have raised as worthy of serious consideration. There is no question the mathematical set–up of the *Pearl* manuscript shows such high intelligence that it can only rightly be called genius. Whether or not it is 'inaccessibly arcane' is another matter. The path through the mathematics is clearly marked with a surfeit of confirming signposts. In fact, it's so clear that even when the territory was completely unfamiliar to me and the mathematical metaphor so strange it may as well have been an alien language, there were enough contextual clues to be able to follow the poet's path, as well as be reasonably certain about a translation of the mathematical metaphor at any particular point.

Still, in favour of John Massey, the rector of Ashton–on–Mersey, is the fact that the *Pearl* poet is extremely well–versed in Scripture. From time to time, he seems to have made a mistake in his Scriptural references but there is not one that cannot be explained as a clue to the design and the overall theological theme. In my view, the 'mistakes' are deliberate.

Another point in favour of a cleric is that the poet read both Hebrew and Greek. If he didn't, he was surely acquainted with someone who did. He is possibly the same John Massey who was rector of nearby Sale, although he may have been simply a relative. The rector of Sale was also the rector of Stockport which is significant in that he was likely acquainted with the previous incumbent, Jordan de Holme. The importance of this is Jordan's suggestive name. Jordan may have been a relative of Ralph Holmes, the so–called 'Green Squire', who was caught up in a coup in Spain and lost his head, quite literally, in a defence of the king. This incident is sometimes seen as impinging on the inspiration for the 'Beheading Game' in *Sir Gawain and the Green Knight*, even though the 'Beheading Game' is an old Celtic literary device. (Haldeen Braddy, *Sir Gawain and Ralph Holmes the Green Knight, Modern Language Notes*, Vol. 67, No. 4., 1952)

Several commentators have suggested it would be profitable to look for a John Massy who lost a daughter named Margery (from Margaret or Marguerite, which both mean *pearl*) around the age of two years. This would fit the theme of *Pearl* and be a punning allusion to its content. While I agree all the references to a *margery pearl* are puns of various kinds, I believe this exercise to be essentially futile. Although there is internal evidence pointing to the possibility that the *Pearl* maiden is an infant who died under the age of two, the child of *Pearl* is, in my view, primarily a reference to the principal criteria for entry to the Kingdom of Heaven. (*And he [Jesus] said: 'Truly I tell you, unless you change and become like little children, you will never enter the kingdom of heaven.'* Matthew 18:3 NIV)

Against this identification of the *Pearl* poet however, John Bowers makes the very valid point that John de Mascy was a very common name at the time. Besides the rectors of Sale and Ashton–on–Mersey, there was John de Mascy of Puddington who supported Henry IV against Henry Percy who was supported by John de Mascy of Tatton. Both of these John de Mascys were slain at the battle of Shrewsbury in 1403. Bowers' most telling point is that these searches have failed to provide any useful insights into the poetry which prompted the research in the first place.

33   Section XV of *Pearl* contains an extra stanza (six instead of five) and various authors have suggestions for removing one of them. Osgood and Gollancz suggested Lines 853–864 should be removed.

34   Ed Condren, quoting Cary Nelson, suggests that this is due to a 'fear of excessive vanity offending God.' *The Numerical Universe of the Gawain–Pearl Poet: Beyond Phi*, p. 17

35   See, for example, W.F. Bolton (ed), *The Middle Ages–Volume 1* of the Penguin History of Literature, Penguin 1993 p. 165 or E.I. Condren, *The Numerical Universe of the Gawain–Pearl Poet: Beyond Phi* University Press of Florida 2002 p.18. Citing Karl Menninger, Condren also suggests the possibility that 101 may go back to a habit of the tribes from Northern Europe of allowing an excess, especially in the setting of limits defined by law. Examples include not only 'a year and a day' but 'one thousand and one nights' and 'a baker's dozen'.

36   Not the 25th as sometimes quoted; this ignores the only even prime, 2.

37   I find no *consistent* evidence in the works of the *Pearl* poet (or Chaucer for that matter) for treating either 1 or 2 as special numbers and plenty of evidence to suggest otherwise.

38   Lawrence P. Schrenk, *God as Monad: The Philosophical Basis of Medieval Theological Numerology* in Robert L. Surles (ed.), *Medieval Numerology: A Book of Essays*, Garland Publishing 1993

39   Outside of Cotton Nero *A.x*, multiples of 101 appear to be fairly rare in medieval poetry, though it is worth noting that Chrétien de Troyes' *Cligés* at 6767 lines is 67 × 101 in length. Going further back, however, FG Lang suggests that, by a stichometric (line) count of the Gospel of Mark, the Greek text should be divided into a prologue and 5 parts, totalling 1515 lines. (quoted by MJJ Menken, *Numerical Literary Techniques in John – The Fourth Evangelist's Use of Numbers of Words and Syllables*, Supplements to Novum Testamentum, Vol LV, E.J. Brill, 1985)

40   Ephesians 1:3–14 is the 202–sentence, while Ephesians 6:12–18 is 101 words.

41   Or 6085 lines, depending on whether the last line of the manuscript, a quotation in Old French possibly referring to the Order of the Garter, was added later or not.

42   EI Condren, *The Numerical Universe of the Gawain–Pearl Poet: Beyond Phi* University Press of Florida 2002 p.88

43   Was Plato right after all? http://news.nationalgeographic.com/news/2003/10/1008_031008_finiteuniverse.html

44   I agree that the 'golden section' is extremely prominent as a mathematical token within the text, but I do not see it as the organising principle. It was one of several arithmetic and geometric features which were equally important to the poet. Condren may be right in suggesting that the work has been laid out originally like a carpet page, but—at the end of the day—it's impossible to be sure this was the case. The exactness of formulation that could be expected from a geometric construction is simply not there. This is not because the poet is sloppy, either in his mathematics or his writing—in fact, he's very meticulous. It's because he was trying to line up at least a dozen numerical metaphors with theological overtones. Sometimes in order to 'fettle together' a matched pair (or even a triplet), a little pushing and shoving was needed and sometimes a stretch.

45 An exemplar of this is the well-known story of Karl Gauss who was known as the 'Prince of Mathematicians'. His teacher set the class a task: add up all the numbers from one to 100. It was supposed to give him several minutes of peace as the class tackled the enormous sum. However, little Karl was at his desk almost immediately with the correct answer. Instead of adding up all the numbers laboriously, Karl had realised that there were 50 pairs that added up to 101 (1 and 100, 2 and 99, 3 and 98, 4 and 97 and so on.) 50 lots of 101 equals 50 × 101 or 5050.

46 There was, of course, always the possibility that the fractional approximation for the golden ratio varied as the spelling of Gawain's name varied: Gawayn, Gawayne, Gawan, Gawen, Gauayn, Gauan, Wawan, Wawen, Wowen, and Wowayn. However, I went to school in the era before the teaching of mathematics involved the modern educational philosophy of obtaining results by consensus and the relativistic concept that there was no single 'correct' answer, so I was reluctant to accept that if there was any other choice.

47 The design is in fact so similar in many respects to the basic concept underlying Leonardo da Vinci's *Homo Quadratus* (or *Vitruvian Man*) that it really does suggest the *Pearl* poet is in the company of that renaissance genius. The figure of the man in Leonardo's sketch is cruciform—suggesting to many people that it is symbolic of Christ—but it is also pentagonal, suggesting to those influenced by *The Da Vinci Code* that it is a symbol of the goddess. The pentagram, however, was throughout medieval times a symbol of the Five Wounds of Christ on the Cross. *Homo Quadratus* thus raises the question of whether Leonardo was proposing a variation of an old Christian tradition. Is that iconic illustration—the cruciform/pentagonal man enclosed in the square and the circle—now found everywhere from tea-towels to t-shirts really as humanistic as has been claimed? Perhaps it is simply a reworking of a question that many Christian intellectuals of the Middle Ages tackled with relish—how best to 'square a circle'.

The possibility that Leonardo was working within this Christian framework without feeling a need to spell it out is made more plausible by the behaviour of his mentor, Luca Pacioli, for whom he illustrated *De Divina Proportione*. In that work, Pacioli gives no indication that he is writing of a very old Christian concept when he describes the 'divine proportion' or 'golden ratio'. Indeed, a number of critics believe he plagiarised the work of his own mentor, the renowned painter Piero della Francesca. However, Pacioli does not explain himself well and I would suggest that the reason for that is that he did not need to. He was simply summarising a very well-known mathematical idea with theological associations which had been explicit for almost a millennium and a half before della Francesca came on the scene. It is unreasonable to expect the 'Divine Proportion' to be attributed to della Francesca when it is, in fact, at least as old as Christianity itself.

48 The poet emphasises the importance of the need for forgiveness with respect to the girdle by using 490 directly later. Gawain puts the girdle over his surcoat on leaving Hautdesert 490 lines from the end of the poem.

49 Moreover, as Condren pointed out, there are 21 stanzas in Fitt I and 34 in Fitt III. These two adjacent 'Fibonacci numbers' demonstrate a well-known approximation for the golden ratio. (While noting Fibonacci numbers, it should also be pointed out that *Purity* is divided in 13 sections and *Patience* into 5.)

50 Significant digits are those figures used to indicate the precision of a number. Leading and trailing zeroes are placeholders which indicate the degree to which the number has been rounded off.

51 Another example of 101, by the way.

52 490 works two ways as the defining number, not simply to produce 101. If, instead of multiplying by it by 0.618, it is divided by the fifth power of 0.618, it yields 5435 which is 651 lines short of the total manuscript. (The fifth power is chosen because of the significance of 5 in *Sir Gawain and the Green Knight*.) Line 651 of *Pearl* refers to the fifth wound of Christ on the Cross and Line 651 of *Sir Gawain and the Green Knight* refers to the fifth of the fivefold pentads of Gawain's shield (among which are the Five Wounds on the Cross.) While these positions are subtly symbolic of forgiveness via Christian theology, a much more overt motif occurs 651 lines from the end of *Sir Gawain and the Green Knight*: Gawain's much-debated confession immediately after he accepts the green girdle.

53 In fact, he thought that the excess lines over the neat thousand (that is 12 extra lines in both *Pearl* and *Purity* and 31 extra lines in each of *Patience* and *Sir Gawain and the Green Knight*) were also musical references. He considered the first two poems to have a key of twelve and the last two to have a key of 10 (the latter because 31 is, in his view of medieval mathematics, the 10th prime number.) While I agree that the excess lines are a musical reference, I do not agree that they designate the 'key' of the poem. I am also reluctant to see 31 as

anything other than the 11th prime number. I am dubious, in fact, about accepting any of Condren's arguments when it comes to primes. I acknowledge that I am a child of modern mathematics, that it's difficult for me to ignore that 2 is a prime number (the only even one there is) and so it seems to me that Condren's prime count is always one out. Condren's prime count highlights Pythagorean features. However, when 2 is accepted as a prime, I believe much more significant features emerge: 17, for instance, is the 7th prime and not the 6th.

54 RR Edwards, *Ratio and Invention, A Study of Medieval Lyric and Narrative*, Vanderbilt University Press, 1989, p6

55 'If, as I say, a man pursues his studies aright [i.e., astronomy, pure numbers, mensuration, solid geometry and harmony] with his mind's eye fixed on their single end...he will receive receive the revelation of a single bond of natural interconnection...' Condren, quoting from Plato's *Epinomis*, on the philosophy underlying the mathematics within the *Pearl* manuscript. (*The Numerical Universe of the Gawain–Pearl Poet: Beyond Phi*, p.13) I would contend the philosophy goes even deeper: the mathematics does not exist for its own sake or even for the sake of the natural union perceived to exist within the subjects of the quadrivium; it exists to make a theological point.

56 Maren Sofie Røstvig, *The Hidden Sense* p20, quoting Gretchen Ludke Finney, *Music: A Book of Knowledge in Renaissance England*, Studies in the Renaissance (1959) p49, quoting Augustine, *De Musica*, tr. RC Taliaferro (Annapolis, 1939) p148

57 CS Lewis, *Christianity and Literature in Religion* and *Modern Literature: Essays in Theory and Criticism*, GB Tennyson & Edward E Ericson Jr (eds.), Eerdmans, 1975. Lewis is specific in his mathematical analogy: '...there is no doubt that this kind of proportion sum – A: B::B: C – is quite freely used in the New Testament where A and B represent the First and Second Persons of the Trinity'.

58 Kay Brainerd Slocum, *Speculum musicae: Jacques de Liège and the Art of Musical Number* in Robert L. Surles (ed.), *Medieval Numerology: A Book of Essays*, Garland Publishing 1993

59 It would therefore unify all of the quadrivium within its bounds: astronomy, music, arithmetic, geometry. This is a very strong argument in its favour.

60 This is obtained as follows: $6000 \times 1.0136432647705078125 \approx 6082$

61 It might seem like a simple coincidence but, when the number five is repeated

so often within this sequence it seems almost obsessive, it is noteworthy that the length of the description is *not* a multiple of five.

62   *Hancque mathematicam dicunt didicisse sorore*
       Moronoe, Mazoe, Gliten, Glitonea, Gliton,
       Tyronoe, Thiten cithara notissima Thiten.

       (It's said she [Morgan] taught mathematics to her sisters Moronoe, Mazoe, Gliten, Glitonea, Gliton, Tyronoe and Thiten, who was a noted player of the cither.)

       Lines 926–928, *Vita Merlini*, Geoffrey of Monmouth

63   By which time, it may be clueing us in to some associations of Morgan, the goddess with a mathematical bent. The references to Morgan end just after line 6018 of the text, which is 17 × 354. This latter number is one we will encounter many times: it is the number of days in the old synodic year. It could well be linked via the golden section to the line numbering of the quote from the *Vita Merlini*, 'Life of Merlin', in the previous footnote.

64   MJJ Menken, *Numerical Literary Techniques in John – The Fourth Evangelist's Use of Numbers of Words and Syllables*, Supplements to Novum Testamentum, Vol LV, E.J. Brill, 1985

65   The name *Sopherim* comes from Hebrew 'caphar', *count, number, enumerate exactly* but also meaning *recount* or *talk*. Names such as Joseph, *God adds*, are related to it.

66   2 Thessalonians 2:2

67   MJJ Menken, *Numerical Literary Techniques in John—The Fourth Evangelist's Use of Numbers of Words and Syllables*, Supplements to Novum Testamentum, Vol LV, E.J. Brill, 1985 quoting Karl Menninger, *Zahlwort und Ziffer, Eine Kulturgeschichte der Zahl*, Göttingen, 1979

68   The publicity blurb for Bartelt's *The Book Around Immanuel—Style and Structure in Isaiah 2–12* is so refreshingly oblivious to modern ideas about numerical literary formats that I simply have to quote it, complete with exclamation marks: there are 496 syllables in Isaiah 5:8–25 and the same number in 9:7–10:4. For the larger macrostructure, Isaiah 2:1–5:30 has 2352 syllables, as does its matching 8:19–12:6! The *Denkschrift* of 6:1–8:18 has 1696 syllables, making a grand total of 6400 syllables, a pattern of one standard 8–

syllable line times 8 times 100. Moreover, the middle line of the *Denkschrift* is 7:14b, and it divides the entire chapters 2–12 exactly into two halves of 3200 syllables each.

69   Quoted by Robert L. Surles, *Medieval Numerology: A Book of Essays*, Garland Publishing 1993, page *x*.

70   Michael Robertson, *Stanzaic Symmetry in Sir Gawain and the Green Knight*, *Speculum, Vol. 57, No. 4 (Oct., 1982)*

71   Eleanor Webster Bulatkin, *Structural Arithmetic Metaphor in the Oxford "Roland"*, Ohio State University Press 1972

72   I am very wary of any statistical analysis of literature. It must be remembered that when it comes to composition, a bad poet mimics, a good poet plays. A bad poet copies, while a good poet is a vanguard of innovation; he experiments with rhyme, rhythm, number, words and form. But even that makes the situation sound simpler than it really is because, as TS Eliot once quipped, 'mature poets steal.' The potential for mistaken attribution or for mistaken non–attribution in these circumstances (that is, without an adequate sample of *known* work by one author) is enormous. The very nature of the situation maximises the chances of identifying a false positive or a false negative solution. These comments apply regardless of the actual quality of the statistical analysis. It can be flawless, but that does not alter my view that, in cases like the *Pearl* manuscript and *St Erkenwald*, it should never have been used in the first place.

73   The poet seems to have been fairly conservative by playing with the inclusion or exclusion of a single line. Ed Condren suggests that Chaucer created the mathematical framework of *Parliament of Fowls* so that stanzas 17 and 70 had in some sense this kind of opt–in, opt–out character to facilitate a geometric construction. (Edward I. Condren, *Chaucer from Prentice to Poet: The Metaphor of Love in Dream Visions and Troilus and Criseyde*, University Press of Florida, 2008) Interesting that such a construction should depend on the exclusion of Stanza 17, the very one that would have been completely indispensable from the *Pearl* poet's perspective.

74   459 lines (17 × 27). 17 is mentioned specifically in the poem when the dreamer climbs a mountain and is able to see for 17 miles in every direction. A little mathematical back–calculation to work out the height of the mountain from the 'distance to the visible horizon' reveals it to be more of a molehill and only a few hundred feet high.

75   604 lines (605 lines is 11 × 55)

76   44 lines (22 × 2): Brian Stone (tr.), *Medieval English Verse*, Penguin Books, 1988

77   44 lines (22 × 2): Brian Stone (tr.), *Medieval English Verse*, Penguin Books, 1988

78   11 lines; Brian Stone (tr.), *Medieval English Verse*, Penguin Books, 1988

79   22 lines; Brian Stone (tr.), *Medieval English Verse*, Penguin Books, 1988

80   44 lines (22 × 2): Brian Stone (tr.), *Medieval English Verse*, Penguin Books, 1988

81   77 lines (11 × 7): Brian Stone (tr.), *Medieval English Verse*, Penguin Books, 1988

82   55 lines (11 × 5): Brian Stone (tr.), *Medieval English Verse*, Penguin Books, 1988. It should be noted in addition that 55 is a Fibonacci number and, as might be expected, this poem shows a sudden break in its regular patterning to accommodate a golden section.

83   66 lines (22 × 3): Brian Stone (tr.), *Medieval English Verse*, Penguin Books, 1988

84   44 lines (22 × 2)

85   22 lines

86   627 lines (11 × 57)

87   The alliterative *Morte Arthure* is a multiple of 11; however if the 11 line prologue is discounted, it is a multiple of 22.

88   This is assuming that number of missing lines at the very end of the manuscript is only 2. However, assuming that 2 is correct, then the call for drinks which divides the poem occurs precisely at a golden section, so is an additional factor in its favour.

89   133 lines (22 × 6 + 1)

90   Condren, in *The Numerical Universe of the Gawain–Pearl Poet: Beyond Phi*, points out the especial fondness of English dramatists in the Renaissance for 33-line soliloquies. He attributes this to a reverence for the traditional life of Christ at 33 years. However, they may descend from the medieval phenomenon described here of multiples of 11 and 22.

91   607 × 11

92   John M. Bowers, *Chaste Marriage: Fashion and Texts at the Court of Richard II*,

Pacific Coast Philology, Vol. 30, No. 1, Pacific Ancient and Modern Language Association, 1995

93 Strictly speaking, for a factor of 11 or 22, the sum of the alternate digits could be the same or else they could differ by 11. For example, 858 is a number we shall encounter in the next chapter. Its alternate digit pairs are 16 and 5: these differ by 11, so 858 can be evenly divided by 11, but it is also even, so it can be evenly divided by 22.

94 Archimedes was born in about 287 BC. The approximation for pi which he calculated by use of polygons was considerably more sophisticated than a simple $^{22}/_7$. He determined that the value of pi was actually less than $^{22}/_7$ but greater than $^{223}/_{71}$.

95 Recapitulation and summary (e.g. Chaucer in *Miller's Tale* and *Reeve's Tale*), supplication for an amendment (e.g. Chaucer calling on Gower in *Troilus*), craving indulgence for deficiencies (e.g. Gower in *Confessio*), the vaunt for glory, the praise of God (e.g. Chaucer in *Troilus.*)

96 As CS Lewis says in *The Discarded Image*, the Natural opening advocated by Geoffrey is to follow the King of Heart's advice and begin at the beginning.

97 The three artificial ways are to begin at the end, to begin in the middle or to begin with a maxim.

98 Ann R. Meyer, *Medieval Allegory and the Building of the New Jerusalem*, D.S. Brewer 2003 quoting Mary Carruthers, *The Poet as Master Builder: Composition and Locational Memory in the Middle Ages*, New Literary History 24, 1993

99 Another circularity is the one pointed out by Jay Schleusener and JA Burrow within the prologue of *Patience*. There, in the discussion of the eight Beatitudes, poverty is 'fettled' to patience—thus the first Beatitude 'blessed are the poor in spirit' is linked to 'blessed are those who are persecuted because of righteousness'.

100 Robert W. Barrett, Jr, *Against All England: Regional Identity and Cheshire Writing, 1195–1656*, ReFormations, University of Notre Dame Press, 2009, pg. 266

101 Maidie Hilmo, *Medieval Images, Icons, and Illustrated English Literary Texts from the Ruthwell Cross to the Ellesmere Chaucer*, Ashgate 2004. Hilmo suggests that the illustrations in the *Pearl* manuscript may be deliberately more 'ordinary' simply to avoid being excised or defaced by the Lollards or their sympathisers.

The circularity she discerns, however, suggests that the illustrations may be more significant to the text than it generally supposed.

102 If truth be told, it was only when I was checking the final draft of this manuscript that I realised I had made a mistake in calculating the size of *Summer Sunday* many years previously. Yes, true—a mathematics teacher who can't add up! It was an incredibly fortuitous error, in retrospect, because I doubt I would ever have seen the nature of the design so obviously anywhere else. However, it is equally true that breakthroughs occur by examining where the pattern does not quite work as expected. This entire book comes out of trying to explain why the golden sections Ed Condren found weren't quite where any medieval fractional approximation would have placed them.

103 A radian simply means that the a circle is turning through an angle which corresponds with an arc on the circumference equal in length to the radius, so a dourad means the circle is turning through an angle which corresponds with an arc on the circumference equal in length to the diameter.

104 His name is mentioned previously but it is not until Line 108 that his character is introduced.

105 JA Burrow, *A Reading of Sir Gawain and the Green Knight*, Routledge and Kegan Paul, 1965, p 56

106 Although this does work very well as an explanation of the decorated initial at the beginning of Line 763, it should also be noted that the three decorated initials are at Line 491, 619 and 763. Line 763 is 144 lines beyond Line 619, again foregrounding the mathematical motif which I suggested might be the 'communion of saints', a thought that gains credibility here as Gawain has just called for the aid of Jesus, Mary and St John.

107 C. Hugh Holman, *Marere$_3$ Mysse in The Pearl*, *Modern Language Notes*, Vol. 66, No. 1, 1951

108 Andrew & Waldron's translation: Andrew, M., Waldron, R., (eds.), *The Poems of the Pearl Manuscript: Pearl, Cleanness, Patience, Sir Gawain and the Green Knight*, University of Exeter Press 2002; p 282

109 RR Edwards, *Ratio and Invention, A Study of Medieval Lyric and Narrative*, Vanderbilt University Press, 1989, p84

110 Fortune was traditionally seen at the time as a servant of the Almighty, but

nevertheless she represents a concept that can easily lose its Christian mooring as the poet would almost certainly have known from Isaiah 65:11.

111 It could be argued that the 66–pattern in the Oxford 'Roland' which Bulatkin dubbed *'démesure'* is an earlier prototype for the dourad. While the *'démesure'* pattern does not depend on the $^7/_{22}$ position, the 66–cycle (or 44 & 22 cycle) is about significant changepoints in the narrative all linked to the tragic flaw of the hero, Roland, whose death occurs in laisse 176, itself a multiple of 22.

112 At this point, perhaps you might incline to the suspicion that, because the poem fails to be a multiple of 22 by just one line but has a dourad anyway, that a line has been lost due to scribal error. This is not the case— as will be evident later.

113 Private correspondence.

114 Richard Amiel McGough, *The Bible Wheel: A Revelation of the Unity of the Divine Bible*, Bible Wheel Book House, 2006, p 21 McGough was prompted to try the same with the books of the Bible and, there being 66 in the Protestant tradition, this naturally led to three circles. McGough looks at many links within these circles but does not consider the 'compass rose' of a single circle. Indeed, on this matter of a circle, McGough points on his website to the work of Vernon Jenkins who found evidence of an astonishingly accurate value of π in Genesis 1:1. See http://homepage.virgin.net/vernon.jenkins/First_Princs.htm#Intro

115 Although this statement by John of Garland is cited as an idiosyncratic view by Robert Edwards in *Ratio and Invention*, it has been an opinion held by authors of fiction even to the present day. As a fiction writer, I hold it myself and did so long before knowing anyone else did. George MacDonald, the nineteenth century fantasist, famously said: *'A man may well himself discover truth in what he wrote; for he was dealing all the time with things that came from thoughts beyond his own.'*

116 Curiously while this indicates a 'poetic circumference' of 10000, it may be inspired by an ancient Roman belief about the *diameter* of the earth. According to Vitruvius (Marcus Vitruvius Pollio)—the Roman architect famous for having inspired Leonardo da Vinci's iconic *Vitruvian Man* (or *Homo Quadratus*), the diameter of the Earth was precisely 10000 miles. (Robert Tavernor, *Smoot's Ear: The Measure of Humanity*, Yale University Press, 32007)

117 Joan Helm, *A Trick That Has Been Lost: Modern Society And Poetry Through The Eyes Of Calliope And Pythagoras In The Poetry of A.D. Hope*, Quadrant, July 2001

118 Attwood, Katrina, *Christian Poetry* in McTurk, Rory, *A Companion to Old Norse-Icelandic Literature and Culture*, Blackwell Publishing, 2005

119 Hill, Thomas Dana, *Number and Pattern in "Lilja"*, The Journal of English and Germanic Philology, Vol. 69, No. 4 (Oct., 1970)

120 'All poets wish they had composed *Lilja*' was the famous proverb which developed in later centuries about the poem with its technical brilliance, its serenity of invocation of God, its dramatic image of the Crucifixion as a cosmic struggle with the World Serpent of Norse mythology caught on the hook of the Cross, its circularity of form (the first and last stanzas) and its revolutionary simplicity of language within the traditional pattern of a formal metrical style.

121 As 'kosmos' means *ornament*, is it reading too much into the mathematics to suggest that forgiveness is the poet's interpretation of cosmos, the ornament worn at God's breast?

122 This calculation uses the standard classical approximations of $^{22}/_{7}$ for π and $^{265}/_{153}$ for $\sqrt{3}$, however, if these are not used but modern irrationals instead, then the required length is 2532 lines, only 1 line different. It also begs the question of whether this same relationship exists within *Pearl*: for a circle with a circumference of 1212 lines, the perimeter of the inscribed triangle would be 1001 lines, so close to 1000 that it seems unlikely the poet would have overlooked this fortuitous situation, assuming of course that he didn't start with 1000 and deliberately contrive 1212 this way.

123 This would be relatively easy to achieve by subtracting 3 lines and reducing the length of *Sir Gawain and the Green Knight* to 2517 lines, the proportions would remain intact. This would mean still mean the total length was divisible by 17. On the other hand, by increasing the length of either *Pearl* or *Purity* or both it would be possible to maintain the overall length of 6086 lines.

124 Perfect numbers have been known since the time of Euclid. They are those numbers whose aliquot factors (not including the number itself) add up to the number. The first perfect number is 6 since its factors are 1, 2, 3 and 6. Those numbers, excluding the 6 itself, add up to 6. The next perfect number is 28 since its factors are 1, 2, 4, 7, 14 and 28. 1 + 2 + 4 + 7 +14 = 28. The third perfect number is 496. This number is highlighted in structure of the prologue and epilogue to John's gospel. Its factors are 1, 2, 4, 8, 16, 31, 62, 124, 248 and 496. 1 + 2 + 4+ 8 + 16 + 31 + 62 + 124 + 248 = 496

125 The notion of 40 as an arithmetic metaphor for testing and purification comes from its association with the forty days of Lent which in turn went back to the forty days Jesus spent fasting in the wilderness before being tempted by the devil and also to the forty years the nation of Israel spent wandering the desert before entering the Promised Land. Mathematically, 40 (like 531) could also be considered as a sum: $3^3 + 3^2 + 3^1 + 3^0 = 40$. A similar formulation is $7^3 + 7^2 + 7^1 + 7^0 = 400$

126 JJ Anderson, *Language and Imagination in the Gawain–poems*, Manchester University Press 2005 *p. 154*

127 Boethian circularity as described in what Ross Arthur calls 'the standard fourteenth century mathematics text, the *De Arithmetica* of Boethius' has its roots in Pythagorean mysticism: any number ending in 5 multiplied by any number ending in 5 will produce a number ending in 5. This is also true for 6, which (because of this property) was regarded as the number of regeneration ($6 \times 6 \times 6$ equalling 216 was considered to be the number of re–incarnation; the number of years after which Pythagoras himself was said to return in a different body.)

128 During the fourteenth century, the Julian calendar (named after Julius Caesar who instituted it) was in operation. As far as the date of Easter was concerned, it was a simple calculation involving the 19–year lunar cycle and the 28–year great solar cycle ($19 \times 28 = 532$). In 1583, under the Gregorian system, Easter Sunday was designated as the first Sunday after the full moon after the spring equinox, except when the full moon falls on the equinox. As a consequence, the date of Easter repeats itself every 5.7 million years or so in the Gregorian calendar as opposed to every 532 years in the Julian system.

129 'These have been the days when accounts had to be settled, days when magistrates paid their visits to outlying parts in order to determine outstanding cases and suits. There is a principle of justice enshrined in this institution: debts and unresolved conflicts must not be allowed to linger on. However complex the case, however difficult to settle the debt, a reckoning has to be made and publicly recorded; for it is one of the oldest legal principles of this country that justice delayed is injustice. Among the provisions that the barons wrested from the extortionate and unjust King John in Magna Carta of 1215, a safeguard for gentry like themselves and hungry peasants alike, was the promise that 'To none will we sell, or deny, or delay right or justice'. Days of assize ensure

openness, assurance and timeliness of justice, justice not sold, not denied, not delayed.' Clines, David J. A. (1998). *On the Way to the Postmodern: Old Testament Essays, 1967–1998*. Continuum International Publishing. p. 801.

130 Jeffrey Burton Russell and Brooks Alexander, *A New History of Witchcraft: Sorcerers, Heretics and Pagans*, Thames & Hudson, 2007

131 Chara Frugoni, tr. William McCuaig, *Inventions of the Middle Ages*, The Folio Society, 2007

132 *http://findarticles.com/p/articles/mi_m0NVC/is_1-2_27/ai_n15389277* Accessed 06/06/08

133 George Sarton & Solomon Gandz, *The Invention of the Decimal Fractions and the Application of the Exponential Calculus by Immanuel Bonfils of Tarascon (c. 1350)*, Isis, Vol. 25, No. 1, The University of Chicago Press 1936

134 Karl Menninger, *Number Words and Number Symbols: A Cultural History of Numbers*, The M.I.T. Press, 1969.

135 Nick Davis, *Recognition of Worth in Pearl and Sir Gawain and the Green Knight* in *The Middle Ages in the North–West*, ed. Tom Scott and Pat Starkey, Leopard's Head Press, Liverpool, 1995

136 There is some difficulty about the stanza's placement as well. Some scholars consider it as starting on Line 510 and some on Line 513. It's hard to tell from a mathematical point of view. 510 is divisible by 17 which makes it tempting to suggest this is right. Moreover, the stanza would then end on Line 512 which happens to be 17×208 lines into the manuscript. In addition the middle line of the stanza in this position, Line 511, is the 3535th line of the manuscript or $5 \times 7 \times 101$ lines. Furthermore, if there are multiple overlaid polygons forming a geometrical pattern in the text, then this is the vertex of both a hexagon and a hexagram overlaid on the last half of the manuscript. (See digression 4 for more details.) It's very tempting to think there is a hexagon given that five out of six vertices line up with very significant structural aspects of the text.

Nonetheless, if the stanza does indeed start on Line 513, then there's still a few mathematical jewels gleaming: there are 512 lines before the irregular stanza and $512 = 2^8$ while 513 itself is anagrammatic of 531 and of 153. Moreover the middle of the stanza is 17 lines from the end, while 513 is 17 lines past Line 496 which is not only a 'perfect number' (one of those which Condren suggests forms the main structure of the poem) and the 31st triangular number but also features exceedingly strongly in the Greek text of the gospel of John, being the number of syllables in his prologue and the number of words in his epilogue. It also features twice as the number of syllables in the chiasmus (or poetry 'ring') surrounding the prophecy of the coming of Immanuel in the Hebrew text of Isaiah.

Thus, from a mathematical and metaphorical point of view, it's too hard to tell. Either position makes a great deal of sense.

137 The modern accommodation of Hinduism and Buddhism to western tastes does not convey the more deeply life–denying aspects of these philosophies in their classical formulation. An exception is perhaps to be found in Robert Pirsig's *Zen and the Art of Motorcycle Maintenance* where the narrator, Phaedrus, is in India listening to a philosophy lecture and hears the same theme for perhaps the fiftieth time: the world is an illusion. He coldly queries whether the bombs dropped on Hiroshima and Nagasaki were illusion. On receiving an affirmative answer, he leaves the classroom and the country. Classical Buddhism and Hinduism regard the western notion of human rights as simply a delusion.

138 And not just one version of re–incarnation either. For the Pythagoreans, 216 was the mathematical motif for rebirth, while for some Buddhists, it is *at least* 547. The number comes from the fact that there are 547 Jakata Tales which relate the rebirth of the Buddha. Some Buddhists take this to mean that the number of rebirths in a cycle must be at least 547. Only 16 more than 531. Whether or not this figured in the choice is simply too hard to tell. It might be argued that 16 is too much of a difference in a numerical literary design— that it should be more exact than that—however in a design so intricate it has perhaps as many as a dozen layers, this is not necessarily the case.

139 We have previously noted the poet's claim that the poem is wrought with 'lel letteres loken' but it should be pointed out that on Line 495 of *Sir Gawain and the Green Knight*, he admits that 'the beginning and the end fold together but seldom'. In the view of Rebecca Gaines and Holly Barbaccia (*Sir Gawain and the Green Knight*, Spark Publishing, 2002, p34) suggest a metaphor which 'compares life to a string or a piece of fabric that doesn't fold together evenly and neatly, recalling the Fates of classical mythology who measure out man's life with a thread.' I would contend that, while the metaphor of the folded fabric is an extremely apt one, that the folds are primarily mathematical in nature and, while they work perfectly 'but seldom', this is one of the few cases where

they do. Generally speaking, however, they are not much out: often just a line or two. In this sense, the idea of fettled lines with 'lel letteres loken' still holds.

140 Perhaps never more true than with regard to mental states which enable the development of new branches of mathematics. Language precedes discovery in a curious way, often it seems that until a new language has been developed to describe that which is about to be found; until then various results are entirely hidden because mathematicians do not have the language to describe them. The revolutionary work of Andre Weil, for instance, was born partly as a result of familiarity with Sanskrit. 'Weil's ability to navigate some of these [mathematical] landscapes where others had failed can be traced back to his early love of ancient languages, especially Sanskrit. He believed that the development of new mathematical ideas went hand in hand with the development of sophisticated forms of language. For Weil it was no surprise that in India the invention of grammar had preceded that of decimal notation and negative numbers, and that the algebra of the Arabs was born out of the sophisticated development of the Arabic language in medieval times.' Marcus du Santoy, *The Music of the Primes: Why an Unsolved Problem in Mathematics Matters*, Fourth Estate 2003

141 Nick Davis, *Recognition of Worth in Pearl and Sir Gawain and the Green Knight* in *The Middle Ages in the North-West*, ed. Tom Scott and Pat Starkey, Leopard's Head Press, Liverpool, 1995

142 Donne uses a similar device in a later age. I have heard it described as 'an admirable conceit'. Such cultural despising seems to miss the point of all poetic impulse.

143 There is a possible disruption in the linkages between the two stanzas on Line 757 to be discussed later.

144 I believe this two line difference can be explained by the poet's predilection for the number 17. Assuming there is one zero for each stanza, then the *wyrd* of Line 249 would really be either 229 or 271 (depending on whether the zeroes have been included or excluded). I think they have been excluded and that this is clearly hinted at by the occurrence of *wyrd* on Line 273. (This, of course, is also 2 lines off where I maintain it should be.) A wheel based on 271 would be 852 lines in length (6 lines different to the wheel based on 273 which, at 858, points to the very middle of the stanza that most commentators see as the 'superfluous' one. Section XV has six stanzas, whereas all others have five, and

this is what pushes the poem over the 100 stanzas to 101.) However, working with 852 and again assuming that a consistent correction has to be made for zeroes, this would point to Line 923. Line 923 is 289 lines from the end or 17^2. Thus the two instances of *wyrd* are close enough to where they should be to indicate (and verify) the existence of the zeroes, but sufficiently different to be 'linked and locked' to entirely different mathematical metaphors—the old synodic year in one case and 17 in the other.

145 Another confirmation of these zeroes is the mysterious decorated initial on Line 1357 of *Purity*. Neither the beginning of a stanza nor the start of a section, Line 1357 is just an ordinary line which is highlighted for no apparent reason. However, if we inspect the digits of the number we discover that 1357 contains the odd numbers in order. This further means that, if *Pearl* is considered to be 1111 lines in length, then Line 1357 of *Purity* is the 2468th line of the manuscript. It now shows the even numbers in order. This alternation from the odd to the even numbers of the decimal system might be sufficient reason to feature it more prominently—thereby clueing a medieval reader into the novel decimal–oriented nature of the text. This would be particularly significant for the poet, given the impression given by the verses of *Pearl* which are 12 lines long. They evoke the 12 lines in an inch and 12 inches in a foot of the common agricultural measures; they do not hint of anything decimal. In fact, they hint of quite the opposite.

Looking further at Line 1357 of *Purity*, we should note that its dourad is Line 857 of *Pearl*, in the middle of the 'superfluous' stanza so many commentators wish to excise. It's therefore a straight link to the two wheels mentioned in the previous footnote. I don't have enough knowledge of Ptolemaic astronomy to analyse this, but so many circles linked together, some within each other, do make me highly suspicious that the orbit of some heavenly body is described in these cycles. They reek of the epicycles of the Ptolemaic system.

146 You may be tempted to think at this point that 353 as the dourad of 1111 is close enough to explain why the Wheel of Fate of 858 lines within *Pearl* falls 354 lines short of the total. This would be to confuse the unknotted length of 1212 with the knotted necklace of 1111 lines which I would prefer to keep separate. Even so, I recognise the possibility that, if this numerical literary design is also labyrinthine in nature (as I suspect it is), it would be possible to cross from one mathematical path to another—and also that the poet may have done so. There are vague clues there is a labyrinth within the text, or

more accurately speaking, a maze since a labyrinth is unicursal. The clues are, however, not obvious enough to be sure. One of them is the size of this Wheel of Fate—at 858 lines, it reflects the length of the labyrinth at Chartres Cathedral which is 858 feet.

147 No mention whatever is made of December 28 (Holy Innocent's Day) during Gawain's stay at Hautdesert. This has puzzled many scholars since the poet is otherwise so meticulous. To be honest, I actually don't think this is a mathematical hint, even though I have just raised the possibility that's that precisely what it is. I think it is more of a literary clue than a numerical one. Time is simply not behaving in an ordinary way at Hautdesert and this is a major hint that Gawain has indeed ridden into an 'Otherworld' when he crossed the water on the way in. A water boundary, a bridge or some other threshold was a signal in literature for a passage into the realms of faery or the supernatural. The poet has already used the 'water boundary' as a literary device in this way when he keeps earth and heaven separate in *Pearl*: the stream is impassable to the Jeweller as he seeks to join the Pearl maiden.

In *Sir Gawain and the Green Knight*, the fact that it is Holy Innocent's Day that disappears from the calendar may well be a clue that the attack will be on the innocent party: Gawain himself.

148 Maarten JJ Menken, *Numerical Literary Techniques in John – The Fourth Evangelist's Use of Numbers of Words and Syllables*, Supplements to Novum Testamentum, Vol LV, E.J. Brill, 1985

149 Joost Smit Sibinga, *The Composition of 1 Cor. 9 and Its Context*, Novum Testamentum, Vol. 40, Fasc. 2, BRILL 1998

150 Or 17 × 17 × 7 × 2 lines

151 Joan Helm, *Eric and Enide: Cosmic Measures in Nature and the Hebrew Heritage* in Robert J. Surles, *Medieval Numerology: A Book of Essays*, Garland Publishing 1993

152 Especially given David Howlett's discovery of 6 verse forms totalling 6666 letters in the Gnomic poems of the Exeter Book.

153 17424 is the number of inches in the perimeter of an acre that is one furlong by one chain (that is, 220 yards by 22 yards—note all these multiples of 22 again!) Helm points out the spiritual significance of the numbers that make up the measure of a standard English acre, demonstrating that it appears to have been

designed with a knowledge of Hebrew cosmology and a cubit of 17.6 inches. (Reminiscent, I might point out of the 1760 yards in one mile.)

154 The design of the city with its twelve jewels, the twelve layers of the foundation of the city wall and its the twelve gates is the link to the motif 'twelve' in the twelve-line stanzas of *Pearl*.

155 Line 1111 of *Purity* which mentions God on his throne, another symbol of the heavenly city, seems to confirm that this is an appropriate interpretation.

156 See, for instance: Ross Arthur, *Medieval Sign Theory and Sir Gawain and the Green Knight*, University of Toronto Press, 1987

157 Line 1111 of *Pearl* mentions 'red gold' (in reference to the seven golden horns of the Lamb) and while I can find no reference to it as a symbol of heaven, it seems likely that there is some connection, since Line 663 of *Sir Gawain and the Green Knight* also mentions 'red gold'. This position is not only 38 × 111 lines from the beginning of the manuscript (38 being significant as one of the golden sections of 100) but is also 1111 lines from Line 1729 of *Purity* which explains 'mene' as a clean and exact number, perhaps punning on 'mean' as in 'golden mean'.

158 JJ Anderson, *Language and Imagination in the Gawain-poems*, Manchester University Press 2005 *p. 102*

159 PM Matarasso, for instance, describes *Queste del Saint Graal* (*Quest of the Holy Grail*) as an anti-romance set within the Romance to end all romances, the Prose Lancelot. It's an aventure (not an adventure, since that does not convey the mystical overtones of the various characters' journeys), chockfull of strange symbols and marvellous motifs, all of which are explained in a spiritual fashion by ensuring the hero of the moment repairs to the nearest hermit or recluse.

160 So reminiscent of one of the earliest chivalric orders: the Order of the Banda, founded by Alfonso XI of Castile in 1330.

161 Obviously ignoring the vague references in Lines 41 & 42 to games over Christmastide that involved tourneys.

162 At least it's never noted that he does. I don't believe it's possible to assume what is not specifically stated. The Gawain we meet in *Sir Gawain and the Green Knight* is definitely not the Gawain of the continental romances.

163 Richard Barber, Juliet Barker, *Tournaments: Jousts, Chivalry and Pageants in the Middle Ages*, The Boydell Press, 2000

164 Richard Barber, Juliet Barker, *Tournaments: Jousts, Chivalry and Pageants in the Middle Ages*, The Boydell Press, 2000

165 When minor corrections have been made due to satellite imaging. The radius of the earth along the polar axis until recently at the time of writing was given as 3800.5 miles (incredible accuracy!), and at the equator as 3813.2 miles (an error of only 0.34% from the medieval standard),

166

> *With his legs at the stride stands the Man in the Moon:*
> *High on a pitchfork is burden he bears.*
> *Great wonder it is that he never falls down,*
> *So sore doth he shudder and shake with his fears.*
> *When freezes the frost he is chilled to the bone:*
> *The <u>thorns</u> rend his clothes all to tatters and tears.*
> *There's not a soul knows if he'll seat him anon,*
> *And only a witch could say what he wears!*
> *Whither away doth he journey, this man*
> *Who sets his one foot the other before?*
>
> *Nobody marked it if ever he ran,*
> *He's the slowest of men that woman e'er bore.*
> *He is gathering sticks in the field, and his plan*
> *Is with their thorns to fasten his door.*
> *With his two–edged axe, he must hew what he can*
> *Or his toil in the field will be idle and sore.*
>
> *This man, just as though he had always been there,*
> *As though in the moon he was born and fed,*
> *Leans on his fork, like any grey friar,—*
> *This crookback sluggard who is so much afraid.*
> *Full many a day hath he been with us here,*
> *And in doing his errand, he hath but ill–sped.*
> *A bundle of briars, he hath hewn him somewhere,*
> *And the bailiff will sue if the price be not paid.*
>
> *Get thy <u>sticks</u> home ere he comes for the fee!*
> *That other foot forward! Step over the stile!*
> *The bailiff a man of high office is he*
> *And shall home to our house and make merry the while.*

We'll pledge him in liquor right warmly; and she,

Our dame douce, beside him shall sit down and smile;

And when he is drunk as a drowned mouse can be,

Quite clean from his purpose we will him beguile.

In vain do I cry! If the dolt be so dull

And deaf, let him go to the Devil, I say!

I yell at him, filling my lungs to the full

But no whit he hurries, still will he stay.

Hop along Hubert, thou lily-faced gull,

Thy bowels, I wot, are all water and whey!

I grind my teeth at the ill–mannered fool

But he'll never come down until dawning of day.

Hop Along Hubert c. 1290

Clifford Bax, *Vintage Verse: An Anthology of Poetry in English,*

Hollis and Carter Ltd., 1945

167 JRR Tolkien seems to have been aware of this: *ereg* is one of his Elvish words for *thorn* or *holly*.

168 The combination of decimals mediated via Arabic learning and Greek thinking in its Pythagorean mode strongly suggests that the court of Troyes had access to *The Epistles* of the 'Brethren of Purity,' a school of Arabic mathematicians who flourished during the tenth century and who may be part of a continuous chain back to the Pythagorean Brotherhood since they write both of Pythagoras as a 'unique sage' and of mathematics as a science centred in the soul. (Dewdney, AK, *A Mathematical Mystery Tour – Discovering the Truth and Beauty of the Cosmos,* John Wiley and Sons, 1999 p. 65 ff.)

169 While the standard definition of an acre is an area 1 chain by 1 furlong, it doesn't have to have a length and width of precisely this size. In fact, an acre can have infinite variation in lengths and widths; but *only* the rectangle which is 1 chain by 1 furlong has a perimeter of 17424 inches. Thus both Chrétien and the *Pearl* poet are using a very particular and specific acre to formulate this mathematical token.

170 'The English acre is the most intriguing of ancient measures because it is virtually equal to a hypothetical geodetic acre defined as one–myriad–millionth

of the square on the terrestrial radius: if both acres are expressed as squares, the difference between the lengths of their sides is less than one part in 1200.' A.E. Berriman, *Historical Metrology: a new analysis of the archeological and the historical evidence relating to weights and measures*, J.M. Dent and Sons, Ltd., 1953, quoted by Joan Helm, *Eric and Enide: Cosmic Measures in Nature and the Hebrew Heritage* in Robert J. Surles, *Medieval Numerology: A Book of Essays*, Garland Publishing 1993

171 On the other hand, Aristotle, who was as important an influence on the Middle Ages as Plato, believed that number and words were discrete, not unified. To be described fully, qualities had to be quantified and compared with well-established standards. Accurate imitation of the natural world represented truth as far as Aristotle was concerned and this was the approach he advocated artists take in *Poetics*. While the difference between Plato and Aristotle is a philosophical chasm, from a practical point of view of constructing a poem using the human body as a model, there's actually not all that much difference. It would be impossible to tell, in my view, using mathematics alone which philosophical view the poet held. Nonetheless, there is the matter of the number 17 to take into account. It should therefore be noted that the *Pearl* poet *may* be a Christian Aristotlean, but he is certainly not a Christian Platonist. Being utterly anti–Pythagorean in his views may in fact make him anti–Platonist.

172 Robert Tavernor, *Smoot's Ear: The Measure of Humanity*, Yale University Press, 2007

173 Consider a square enclosed by a circle: the ratio of the circumference to the perimeter is approximately 1.111 (using 22/7 for π) and thus is suggestive of the mathematical grammar of the City of God. Leonardo's *Vitruvian Man* thus suggests the 'body as city' and begs the question of whether he was applying a long–standing literary and mathematical combination to art (which, contrary to received wisdom, had never disappeared to be resuscitated during the Renaissance, but which had been used continually throughout the Middle Ages.)

174 Note that 3168 is missing and this is precisely the number with both numeric and numerological overtones. It is this sort of absence that leads me to the conclusion that the poet avoided almost all numbers that smacked of numerology, even if there were other perfectly legitimate reasons for using them.

175 EI Condren, *The Numerical Universe of the Gawain–Pearl Poet: Beyond Phi* University Press of Florida 2002, pp 69–73

176 The Dreamer does not make it over the threshold, however. The poet's subtlety is perhaps most in evidence at this point. As Condren points out, the Dreamer's speech at this moment closely resembles that of the Pharisee in the parable of the Pharisee and the Publican. While the poet has not explicitly referenced this particular parable, Condren notes there is a strange arrogance in what the Dreamer says. Perhaps a medieval audience would have got the point we tend to miss. Jesus makes the comment on this parable that, *'whoever does not accept the Kingdom of Heaven as a little child will not enter it.'* (Luke 18:17) Herein lies the reason I think it is futile to look for a Margery Massy who died at less than two years of age. I believe the poet was writing of the archetypal Child who has the attitude necessary to enter the Kingdom of God and, through a mystical union with the Son, brings heaven to earth. Such a Child can be 7, 17, 70 or any age above, below or in between.

177 It's difficult to be sure, one way or another, about the poet's use of 12345679. As noted previously, decorated initials occur within Fitt II of *Sir Gawain and the Green Knight* at Lines 491, 619 and 763. Line 491 is immediately after 490, the mathematical motif of forgiveness—it makes sense to make such an important metaphor as reconciliation. Line 619 is immediately after 618, the significant digits of the golden ratio. Again it makes sense to make such an important metaphor as troth: truth and covenant faithfulness. 763 may be the changepoint of the main wheel within the poem (which, unusually, is smaller than the poem itself or its divisions). It has been previously pointed out that 763 is 144 beyond 619 and this alone may be a sufficient reason to highlight it. However, it is also 618 × 1.2345679—and it may be the combination of these last two mathematical features and the poet's desire to ensure that all the mathematics was linked and locked which produced the incentive to move the changepoint of the poem from Line 805, where it would naturally lie, to Line 763.

178 *Language and Imagination in the Gawain–Poems*, Manchester University Press, 2005, p65

179 This appears to go back even further, perhaps to Rabbi Simon of the first century. This same passage is used in Jewish catechetical texts of the present day: *When God was about to create Adam, the angels formed groups, some saying, 'Create him,' and others saying, 'Don't do it.' This is why it is written in Psalm 85:11, 'Mercy [love] and truth fought, justice and peace went into combat.' Mercy [Love] said, 'Create him, because he will practise mercy.' Truth said, 'Don't create him. He will practise lies.' Justice said, 'Do it, because he will be just.' Peace*

opposed it, saying, 'Don't, because he will bring strife.' What did God do? He threw Truth down to the earth as it says in Daniel 8:12. 'Lord of the angels [Master of the universe],' the angels dared ask, 'why did you despise [your seal and] your first bodyguard, Truth? Let Truth arise from the earth.' Thus it is written in Psalm 85:12, 'Let truth spring up from the earth.' (From Hope Traver, *The Four Daughters of God, A Study of the Versions of this Allegory with Especial Reference to those in Latin, French and English*, The John C Winston Co, 1907, with minor bracketed variations from Kerry M. Olitzky, Ronald H. Isaacs, *I Believe: The Thirteen Principles of Faith: A Confirmation Textbook*, KTAV Publishing, 2003)

Traver makes the very significant note that this Midrashic story can only have arisen after Aramaic supplanted Hebrew, since it interprets Psalm 85 on the basis that there are two words of double meaning: 'meet' means 'fight' and 'kiss' is taken as 'arm one's self'.

180 Made famous in the sermons of Hugh of St Victor (who appears to have used a Jewish source) and Bernard of Clairvaux. (Hope Traver, *The Four Daughters of God, A Study of the Versions of this Allegory with Especial Reference to those in Latin, French and English*, The John C Winston Co, 1907)

181 Anne Eggebroten cites a significant English translation of the concept in the thirteenth century. (*Sawles Warde: A Retelling of De Anima for a Female Audience, Mediaevalia,* Vol. 10, 1984) Traver points out that, in the fourteenth century, William Langland made use of the idea in *Piers Plowman.*

182 In *Patience*, for instance, Jonah becomes arrogant and presumptuous when it becomes evident that God's extreme mercy will prevail, rather than his justice. Burrow points out regarding *Sir Gawain and the Green Knight* that it deals with a superlative hero in a superlative test. (JA Burrow, *A Reading of Sir Gawain and the Green Knight*, Routledge and Kegan Paul, 1965, p. 163) He also points out that Gawain conspicuously does not fail the test of 'cleanness' (sexual purity). As he shouldn't—since, if Gawain's story is the unfolding of the argument of Lady Truth, then the issue is primarily *trawthe*, not righteousness.

183 JJ Anderson makes the point that the poem is not logical but works by repeated and emphatic assertion. It's basically rhetoric. The poet's illustrative stories do not expand the argument or enlarge any perspective, they simply re-inforce the point. 'The poem is in the business of rhetoric rather than exposition, and it may be regarded as, in essence, one long rhetorical flourish based on the opening proposition.' *Language and Imagination in the Gawain-Poems,*

Manchester University Press, 2005, p.84

184 None of the poems contained a balanced argument, in and of themselves. The Dreamer, for instance, in *Pearl* (Lines 591–596) appeals to the psalms in remonstrating with the maiden about God's justice. He ignores the reference to God's mercy in the same psalm. A similar selective use of Scripture occurs in *Purity* during the story of Lot's escape from Sodom. All reference to God's acceptance and compassion (eg Genesis 19:21) is bypassed. It's all about righteousness and justice. This is perfectly in keeping, however, with a rhetorical argument on a specific theme from each of the Four Daughters in turn. In such a debate, there would be a narrow focus on what fits the argument as well as a tendency to completely ignore what doesn't. The logic is therefore not always persuasive, of course. It's not about logic—it's about rhetoric! The finer points of the debate (at least in *Purity*) are thus lost in a relentless hammering and in a markedly skewed view. Nonetheless there are moments when the rhetoric of a poem tends towards a greater balance. JJ Anderson makes the point in regard to the Pearl maiden's argument about God's mercy determining an ultimate heavenly reward that this squares with what God tells Jonah in Lines 520–523 of *Patience*: few would prosper if he operated justice without mercy. Ever so briefly, there is a recognition within *Pearl* that there is more than one side to the argument. Anderson, however, also points out the issues are so polarised in *Cleanness* that God actually appears as hostile and remorseless towards sin: there are no degrees of transgression or of punishment—virtue is virtue and sin is sin and the wages of sin is death. No mercy, no grace. In Anderson's view, it's a relief to get to *Patience*. (*Language and Imagination in the Gawain–Poems*, Manchester University Press, 2005, p86ff)

185 The motif of 'The Four Daughters of God' is sometimes referred to as 'The Parliament in Heaven' and also as 'The Debate of the Virtues' as well as 'The Reconciliation of the Heavenly Graces'. As a widespread theme in sermons, poetry or drama, it lasted from the eleventh century into the seventeenth. Faint traces of it can be found in the early twentieth century in such hymns as the so–called 'Love Song of the Welsh Revival' by Gwilym Hiraethog, where the last two lines seem to maintain the old allusion to the Heavenly Graces:

> *Here is love, vast as the ocean,*
> *Loving–kindness as the flood,*
> *When the Prince of Life, our Ransom,*
> *Shed for us His precious blood.*
> *Who His love will not remember?*

Who can cease to sing His praise?
He can never be forgotten,
Throughout heav'n's eternal days.

On the mount of crucifixion,
Fountains opened deep and wide;
Through the floodgates of God's mercy
Flowed a vast and gracious tide.
Grace and love, like mighty rivers,
Poured incessant from above,
And heav'n's peace and perfect justice
Kissed a guilty world in love.

Prior to this, however, Hope Traver and others suggest that allusions to it can be found in the courtroom scene of Shakespeare's *Merchant of Venice* as well as in Milton's *Paradise Lost*. Milton apparently also used it in *Morning of Christ's Nativity*:

Yea, Truth and Justice then
Will down return to men,
Orbed in a rainbow, and, like glories wearing,
Mercy will sit between
Throned, in celestial sheen (141–5)

http://www.thomondgate.net/doc/companion/Companion.htm#four
accessed May 2011

186 Burrow points out that truth is held very lightly today (although he does not mention the passing of the notion of absolute truth in favour of what is true for the individual, it is worth noting its impact on a present–day reader). Consequently this makes it difficult to understand the earnestness of the games in *Sir Gawain and the Green Knight* and the seriousness Gawain has for his word, even when that word is given in nothing more than a game. Gawain does not expect any mitigating conditions to come into play, even when he finds he has been tricked, and even when he belatedly discovers potentially–fatal hidden rules.

The idea that a man would die to keep his word whatever the circumstances does not sit easily with modern sensibilities. Burrows suggests 'good faith', 'trustworthiness' or 'fidelity to contract' as better translations of *trawthe*.

(JA Burrow, *A Reading of Sir Gawain and the Green Knight*, Routledge & Kegan Paul 1965, p24) However, even these meanings fall far short of the high-minded integrity the word should convey. Burrow compares Chaucer's story of Dorigen's playful statement during after-dinner dancing that she would accept the unwelcome advances of Aurelius only if he removes all the rocks from the coast of Brittany. Aurelius enlists the help of a magician and holds Dorigen to her word. Dorigen chooses not to plead extenuating circumstances, even though she could, but displays her own *trawthe*. To keep to a word, even one given in jest to get rid of an unpleasant suitor, is the very nature of *trawthe*: honour, integrity and faithfulness to God and self are all tied up in the notion.

187 JRR Tolkien, *Sir Gawain and the Green Knight* in *The Monsters and the Critics and Other Essays*, Harper Collins 1983, pg 75

188 This, as Tolkien points out, is only really evident on re-reading. Tolkien makes the statement (JRR Tolkien, *Sir Gawain and the Green Knight* in *The Monsters and the Critics and Other Essays*, Harper Collins 1983, pg 79) that *Sir Gawain and the Green Knight* is not 'a mathematical allegory'—yet there is a sense in which that is exactly what it is. But because of its depth—its union of trivium, quadrivium and theology—it also transcends them all.

189 JA Burrow, *A Reading of Sir Gawain and the Green Knight*, Routledge & Kegan Paul, 1965, pg 139

190 Some critics may suggest that the lofty serenity of the poem is spoiled more than once by the attitude of the Pearl maiden towards the Dreamer. She seems at times to be quite harsh, treating the Dreamer without the gentleness that some commentators expect should be her response. Spiritual peace has however never been associated with a sweet cloying sympathy. The old blessing '*May the peace of Christ disturb you*' testifies to its robust nature and its natural alliance with truth. The Pearl maiden fits the character of Lady Peace beautifully.

191 It may be argued, against this, that the margins are not even. However 5:8 is one of the simplest and most well known fractional approximations for the golden ratio. Why insist on 'even' when the possibility of including another golden ratio formulation exists?

192 Reciprocals are numbers like ¾ and $^4/_3$ or $^2/_3$ and $^3/_2$ or ½ and 2. For any number, x, the reciprocal is its inverse power: $(x)^{-1}$

193 Herodotus was allegedly told by Egyptians priests that the pyramids were constructed so that the area of each face was equal to the square of its height. Unless this is an incorrect translation (which has in fact been suggested by Markowsky), this automatically means that the golden ratio was a planned feature of the structure. The difficulty with Markowsky's criticism is that while the angles do not work exactly, nonetheless they are only a few degrees out of alignment for such a design. It's possible that construction constraints are responsible for the discrepancy, though there is much disagreement over interpretation. Roger (Herz-)Fischler, who suggests that the quotation from Herodutus is actually non-existent, nonetheless agrees that the structure is very close to requirements, suggesting however that it is also close to $2/\pi$ (or 0.637) and some common fractions as well. Fischler argues that even Pacioli's 'Divine Proportion' should be treated with caution since, in practical rather than theoretical circumstances, Pacioli advocated use of simple proportions. (Roger Fischler, *On the Application of the Golden Ratio in the Visual Arts*, Leonardo, Vol 14, No. 1, Pergamon Press, 1981.) If we were, however, to take this extreme purist point of view, the golden ratio cannot exist anywhere outside the mind. Strictly that's true—it cannot exist in the real world, it can only be approximated—and this may well be Fischler's point. Although he made a caveat regarding an artist's intention, he savaged the list of usual suspects when it comes to artistic expressions using the golden ratio, including amongst those he believed did not use it Seurat, Juan Gris and Le Corbusier. However, an approximation of the golden ratio by use of simple proportions is, in my view, a valid artistic approach. Why should we insist that artists get out a straight rule and compass to ensure 'perfect' accuracy? It's impossible in practice. And, after all, even the well-known and oft-quoted decimal fraction 0.618 is not actually the golden ratio—it is merely a practical approximation of an infinite decimal.

On this issue of art, it should be pointed out the predominant schools of landscape painting during the nineteenth and twentieth centuries were completely baffling to aboriginal artists of the Australian desert who specialised in dot pictures. (But certainly nothing like the dots of the Impressionist Masters.) These desert artists dubbed the obsession they saw in such work as the 'two-thirds sky' syndrome. The roots of this syndrome, however, are in the classical and aesthetic ideals of proportion emanating out of ancient Greece: one-third and two-thirds are fractions easily judged by eye. Two-thirds approximates the well-known fraction, 0.618, itself a decimal truncation of the golden ratio.

194 *The Curves of Life* listed thousands of examples of flora and fauna which have the tendency to exhibit 'phi' as a growth factor. Since the time it was published, an enormous set of additions have been made to the listing. In addition, if we consider the golden ratio as well as phi, then we'd have to include some strange bedfellows in the mix: the Cassini Division in Saturn's rings, the signal of the pulsar in the Crab Nebula, the sunspot cycle, the branching of lightning, snowflake fractals, the shape of a falcon's gyre, the ratio of drones to workers in an ideal beehive, the spiral of a whirlwind seen from space, the proportion of the length to the width of a chromosome, the orbitals of the hydrogen atom, the transition between Newtonian mechanics and relativistic physics—and even the shape of the universe itself! It appears that Plato was on the right track: the best model of the universe appears to really be a dodecahedron. This is a solid which has 12 faces shaped like a pentagon, to which the golden ratio is intrinsic.

http://physicsworld.com/cws/article/news/2003/oct/08/is-the-universe-a-dodecahedron

Of course, if we measure any of these naturally occurring objects, we discover they aren't perfect examples of the golden ratio. (In some cases, the measuring itself interferes with the result.) But the *tendency* towards the golden ratio is clearly there.

Whether it's a pinecone or a pineapple, the iridescent barbels of a peacock's tail, the stripes of a zebra, the marking of a butterfly, the florets of broccoli, the seedhead of a sunflower, the star-shape in the centre of a pear or apple, the coiled tail of a chameleon or a seahorse, the feathers of a lyre bird, the fiddlehead of a new fern leaf, the cross-section of a fossil ammonite, the curve of a comet's tail, a dophin's fin, an elephant's tusk, a ram's horn, a cat's claw or a bird's beak, the golden ratio is lurking close at hand. And on and on and on. It's everywhere. Even the answer to that hoary old question, 'Why is a banana bent?' has to do with the golden ratio.

A word of caution: certain eminent mathematicians have disagreed that the golden ratio is as ubiquitous as it is, singling out the nautilus shell as a perfect example of an iconic golden ratio emblem which doesn't actually show the golden ratio. They have pointed out that the nautilus does not grow in regular angles corresponding to 222° or 138° (the so-called 'golden' angles.) I agree, it doesn't. However, my advice is: check the arc lengths of the chambers, gentlemen.

195   This mistaken attribution to Kepler results from a misreading of *Mysterium cosmographicum*—'Geometry has two great treasures: one is the theorem of Pythagoras; the other, the division of a line into extreme and mean ratio. The first we may compare to a measure of gold; the second we may name a precious jewel.' Kepler in fact appears to have used the term 'Divine Proportion'.

196   The irrational numbers *cannot* be expressed as either finite decimals or as repeating decimals. Roger Herz–Fischler whose work on the history of the golden section is by far the most comprehensive has admitted an inability to find this reference to the 'sectio aurea' by Campanus. However, it seems from the following late nineteenth century quote that it was known at the time but that the exact reference has since been lost: '*The subject of the golden section is not discussed by Euclid; he had a knowledge of it and mentions it in his works, though not under this name. The name, though it has an ancient ring is not found in ancient literature. Aristotle does not mention the subject, but it is claimed that in his philosophical reasoning, there rules the principle of the golden section; i.e.: the relation of the whole to the part and the parts to each other. His ideas were not carried out by the ancient philosophers but they were the source of much of the speculation in mediaeval times, when mathematical and philosophical thought were closely allied. One writer, John Campanus of Novara, thought that the principle of the golden section descended from the gods.*' Emma C. Ackermann, *The Golden Section*, The American Mathematical Monthly, Vol. 2, No. 9/10, Mathematical Association of America, 1895

197   Roger Herz–Fischler, *A Mathematical History of the Golden Number*, Dover, 1998

198   This is not to suggest that 'gold' as a word to mark the golden ratio did not arise in antiquity, merely that it is difficult to accumulate a critical mass of evidence to support such a case.

199   This is another reason why remarks such as 'I found 23 petals on this daisy' simply demonstrate ignorance, not worthwhile criticism. 23 is not a ratio: it is a simple integer.

By the way, in order to find the golden ratio in any number sequence, there is only one restriction which needs to be considered. This is that, in choosing the two numbers for the initial pair, two zeroes cannot be chosen. One zero is fine. Depending on the choice of initial pair, somewhere between the first and thirteenth iteration (at most), the ratio of one term to the one before it will be very close to phi.

200 Or exactly $(0.5 \times \sqrt{5} - 0.5)^2$ of the way along the line.

201 Roger Herz–Fischler, *A Mathematical History of the Golden Number*, Dover, 1998, p.166

202 Also called Solomon's Sign, Solomon's Shield, Solomon's Signet, Solomon's Star, Wise Men's Star, Star of Bethlehem, Three King's Star, Star in the East, Star of Jesse, Star of Jacob, Star of Heavenly Wisdom, Star of the Sea, triple triangle, Barn Star, Pentangle, Pentacle, Pentancle, Pentagle, Pentalpha (from the fact it looks like 5 A's), Drudenfuss (usually translated "druid's foot" from *druttenfuss* or sometimes "druid's hedge" and said to be painted on the sandals of druids or to be used by druids as a hedge of protection against the demonic; but see the correction at http://freemasonry.bcy.ca/anti-masonry/pentagram.html (accessed 19/06/08) to *trutenvuoz*, said to be related to the footprint of a nightmare–inducing witch.) Also regarded as integral to the 'Seal of God' by the English magician, Dr. John Dee. Used as a Hebrew symbol of truth. Not known to be associated with black magic or the demonic (but rather seen as a protection against it) until the middle of the nineteenth century.

203 Suggestions that Éliphas Levi was simply recording an older unwritten tradition are, in my view, simply untenable. Logic alone dictates that the pentagram has long been recognised in the West as a symbol of God, just as the poet indicates in *Sir Gawain and the Green Knight*. Whether it does indeed go back to the time of Solomon is debatable, however it was in use in Israel about 2000 years ago (as demonstrated in the still–existing sculptured stonework of the synagogue of Capernaum, where Jesus himself may have preached) and is said to have been a symbol of Jerusalem. Its time–honoured use by magicians such as John Dee has always seen it acting as a shield against demons. Clearly a sorcerer who chalks a pentagram on the floor to protect himself from the devil implicitly recognises that the pentagram symbolises a power greater than the devil. Mephistopheles may have been able to gain access to Faust's room because the pentagram chalked on the threshold was missing a corner but the device cannot logically symbolise the devil or black magic, since either would also be an invitation to enter, not a barrier.

204 MJJ. Menken, *Numerical Literary Techniques in John – The Fourth Evangelist's Use of Numbers of Words and Syllables*, Supplements to Novum Testamentum, Vol LV, E.J. Brill, 1985 quoting Karl Menninger, *Zahlwort und Ziffer, Eine Kulturgeschichte der Zahl*, Göttingen, 1979

205 Protagoras, a pre–Socratic Greek philosopher, quoted by Plato in *Theaetetus.*

206 For a lot of people, it isn't, even though it comes close. Tavenor asserts that the renowned twentieth century architect, Le Corbusier, nevertheless designed his 'Modular Man' with the premise that it is (though this is disputed by Herz–Fischler who suggests Le Corbusier's work is based on equilateral triangles.)

207 He also indicates that the translators of the Septuagint were not so careful in adhering to the mathematical design in Genesis, nor in making the distinction between different Hebrew titles for God. (David Howlett, *British Books in Biblical Style*, Four Courts Press, 1997 pg 47ff)

208 Had he gone further he might have noticed that Matthew actually used 'Father' 44 times and 17 is $^1/_\varphi{}^2$ of 44, which also constitutes a golden section.

209 Otto G. von Simson, *The Gothic Cathedral: Design and Meaning*, The Journal of the Society of Architectural Historians, vol. 11, no. 3, 1952

210 Jeoragldean McClain, *Observations on the Geometric Design of Saint-Yved at Braine*, Zeitschrift für Kunstgeschichte, 49 Bd., H. 1, Deutscher Kunstverlag GmbH Munchen Berlin 1986

211 Michael T. Davis and Linda Elaine Neagley, *Mechanics and Meaning: Plan Design at Saint-Urbain, Troyes and Saint-Ouen, Rouen,* Gesta, Vol. 39, No. 2, *Robert Branner and the Gothic*, International Center of Medieval Art 2000

212 MF Hearn *Conferences Celebrating the Nine Hundredth Anniversary of the Beginning of Durham Cathedral: "Engineering a Cathedral," and "Anglo-Norman Durham, 1093-1993"* The Journal of the Society of Architectural Historians, Vol. 53, No. 4 Society of Architectural Historians 1994

213 Robert D. Stevick, *The 4×3 Crosses in the Lindisfarne and Lichfield Gospels*, Gesta, Vol. 25, No. 2 International Center of Medieval Art 1986

214 François Bucher, *Medieval Architectural Design Methods 800–1560*, Gesta, vol.11 no. 2, International Center of Medieval Art 1972

215 √3 had been approximated since the time of Archimedes as $^{265}/_{153}$ and the 153 aspect of this (known as 'the number of the fish') was used repeatedly by the *Pearl* poet.

216 Canons of Page Construction (accessed 12/08/08) http://www.nationmaster. com/encyclopedia/Canons–of–page–construction

217 Robert Lucas mentioned by Eleanor Webster Bulatkin in *Structural Arithmetic Metaphor in the Oxford "Roland"*, Ohio State University Press 1972

218 SK Heninger, Jr., Thomas Elwood Hart, Daniel Laferriere, RG Peterson, *Measure and Symmetry in Literature* Vol. 92, No. 1 Modern Language Association 1977

219 George E. Duckworth, *Structural Patterns and Proportions in Virgil's Aeneid*, The University of Michigan Press, 1962. George Markowsky has correctly criticised certain mathematical aspects of Duckworth's analysis. (George Markowsky, *Misconceptions about the Golden Ratio*, The College Mathematics Journal, Vol. 23, No. 1, 1992) However, his criticisms do not invalidate Duckworth's hypothesis in toto, as some commentators suggest. They call into question *some* of the conclusions but by no means all. The end result is that they simply reduce the number of structural examples of the golden ratio to be found in the *Aeneid*—but they do not sweep all of them away.

Markowsky raises some extremely important and unquestionably valid issues with respect to the golden ratio across various architectural, artistic and literary areas; he is right to be skeptical in some instances but his supporters tend to proceed to the opposite extreme and demand a precision Markowsky himself does not. The situation in literary analysis is far more complex than either Markowsky or his supporters allow: generally speaking, I regard his allowance of 2% as far too liberal and the 'no fractional approximation' advocates as far too extreme. Zero tolerance would be a reasonable position to take in literary analysis if and only if the golden section constituted the sole mathematical constraint. However this is rarely the case.

220 Edward I. Condren, Chaucer from Prentice to Poet: The Metaphor of Love in Dream Visions and Troilus and Criseyde, University Press of Florida, 2008

221 Theodore Silverstein, *Sir Gawain and the Green Knight*, A New Critical Edition, University of Chicago Press 1974, p130.

222 I was intrigued by the difficulty that Condren noted with regard to the *Pearl* poet's use of the golden ratio: he noted it was impossible to work out what fractional approximation was being used. That, to my mathematical mind, seemed incredibly unlikely. The design should reveal a number sequence or pattern which would reveal the fraction immediately. However, as I looked into the matter, it was quickly evident that Condren was right. It was impossible! Moreover, he was also right in his suggestion that error didn't accumulate over repeated use. In fact, sometimes the error margin reduced. What was going on?

Why didn't the poet (and his contemporaries) use $^{618}/_{1000}$ all the time instead of $^{62}/_{101}$ or $^{13}/_{21}$ or $^{21}/_{34}$ or $^{50}/_{81}$? The possibility that mathematical choice was as whimsical as the choices for the spelling of Gawain's name quickly occurred to me—however, the fact is that the choice of spelling for Gawain was not really arbitrary but depended on the constraints of alliteration and rhythm. So what constraints dictated the choice of fraction? It appears that $^{62}/_{101}$ is used when the presiding metaphor is the Music of the Spheres and that $^{50}/_{81}$ is used when the presiding metaphor is the heavenly City of God, since, extremely fortuitously, $(1.111...)^2 \div 2 = {}^{50}/_{81}$ or approximately 0.617. However, at the end of the day, I believe that the *Pearl* poet was using 0.618 for the golden ratio. Whenever there is a discrepancy, it is necessary to look for other motifs in the mix.

223 The appropriateness of Venus at a golden section has a double sense: it refers of course to the goddess of love, so apt as a reference to courtly love, but the golden section is also appropriate to Venus, the planet. Inferior conjunctions of the planet Venus have been known to form an almost perfect pentagram since Babylonian times. The possibility that there is a planetary reference here is enhanced by the fact that Venus is curiously invoked in the poem on Line 113: the period in years between transits of Venus at the time.

224 There may be an exception that proves this rule—the *Pearl* poet, as we shall see, may have wanted the whole of the last 12 lines of *Pearl* to be golden-sectioned, but *Parliament of Fowls* doesn't qualify in that respect as having a small tail of lines over the even hundred. Bede's left-hand numbers simply aren't there.

225 The golden sections of these asymmetric parts show two of the usual suspects when it comes to keywords: *gold* and *peace*.

226 This is to assert that there is no line missing from the poem: that it really is supposed to be 99 stanzas of seven lines plus 1 stanza of six lines and that it is a mistake to think that 100 rime royal stanzas were ever intended.

227 Perhaps Chaucer's mathematics is deliberately ambiguous. He knew Guillaume de Mauchat, but it's possible that nearer to home he was aware of, and concerned about, a poet of 'fructous intelligence' who would condemn any trace of a pro-Pythagorean outlook. From the fact *St. Erkenwald* was almost certainly written in 1386 and *Parliament of Fowls* in 1384 or 1385, it's clear that two highly mathematical poets were working at the same time. The *St. Erkenwald* poet is a firebrand of anti-Pythagorean attitude. Chaucer's arithmetic which appears to flirt with Pythagoreanism without indicating if

it has committed itself wouldn't have got past him without critical comment.

228 JB Trapp (ed.), *The Oxford Anthology of English Literature: Medieval English Literature*, Oxford University Press, 1973

229 I have chosen *paz* for its diverse associations. It's Spanish for both *gold* and *peace*. Chaucer uses pes, *peace*, at the *tau* position of *Parlement of Foules* (in fact he uses 'gold' at the *paz* position, reflecting the romances of Chrétien de Troyes) while the *Pearl* poet uses both 'peace' and 'peas' which suggests some serious punning. Spanish paz, *gold*, is related to Modern Hebrew paz, *gold* from Old Hebrew pharez, *fine gold*. (It might also be connected to Uphaz, an unknown place renowned for fine gold mentioned in Jeremiah 10:9 and Daniel 10:5.) Old Hebrew however also features phares: *divide, division, ratio*. The circumstantial evidence points towards a word like *paz* being a term for one of the golden sections. Even if there isn't such a term, the connotations are suggestive enough to recommend it as a term for the 0.382 position and to thereby reduce the awkwardness of terminology considerably.

230 Ross G. Arthur, *Medieval Sign Theory and Sir Gawain and the Green Knight*, University of Toronto Press, 1987

231 John Gardner, *The Alliterative Morte Arthure, The Owl and the Nightingale and Five Other Middle English Poems in a Modernized Version with Comments on the Poems and Notes*, Southern Illinois University Press, 1971

232 This structurally embedded 17 is an important reason why I think this is extremely likely to be the *Pearl* poet's work.

233 Assuming that the $^1/_\tau{}^3$ construction means anything at all (and this is a very big assumption indeed,) then at 31 or 32, it may be indicative of the lifetime of Richard II who reigned 22 years and 99 days. Poetically speaking, this multiple of 11 may simply have cried out for verses on Fortune's Wheel and the fate of kings.

234 Translated JRR Tolkien in *Sir Gawain and the Green Knight, Pearl and Sir Orfeo*, Unwin Paperbacks 1975

235 The number of lines goes up by the next counting number. The number of lines also belongs to that set of numbers so beloved of cultures in the ancient world: the triangular numbers. They include the tetrakys of the Pythagoreans.

As the sequence gets higher and higher, the form of the 'rose' becomes more apparent.

In mathematics, a compass rose is made simply by spacing any number of points evenly around a circle and then joining every point to every other point. If, for example, there is only one point on the circle, there are no joining lines. If there are two points, there is one joining line. If three, three joining lines. If four points, six joining lines. If five, then ten joining lines. The pattern is quickly evident:

Number of points	Number of lines
1	0
2	1
3	3
4	6
5	10
6	15
7	21
8	28
9	36

236 Why this sequence doesn't progress one more step to the ultimate conclusion of 'quadruply triangular' by noting that 3 is the 2nd triangular number, I don't know. It's one of the philosophical—or perhaps theological—mysteries of mathematics. Invoking 1 as the sacred monad simply doesn't work as an explanation, because that makes 231 the 20th triangular number and 20 doesn't fit the bill for the progressive declension.

237 It may be coincidence (and it probably is, but I wouldn't count too definitely on it) but I'd like to point out that the dourad of the Wheel of Fate formulation in *Sir Gawain and the Green Knight* when divided by the square of *tau* and rounded to the nearest whole number is divisible by 17. $^{805}/\tau^2 \approx 2 \times 62 \times 17$. Or to put it another way: $\tau^3 \times 100 \times 2 \times 17 \approx 805$. Moreover, the metaphor for forgiveness times the diatonic comma divided by tau is also 805. That is, $490 \times 1.013 \div \tau \approx 805$. Whether 805 is fettled in this way is an open question: the evidence is not as clear-cut as elsewhere.

238 He actually seems to have been the third or fourth bishop of London, not the first.

239 El Condren, *The Numerical Universe of the Gawain – Pearl Poet: Beyond Phi*, pg 84ff

240 This naturally suggests that *Sir Gawain and the Green Knight* was in preparation in 1386. However, I don't think it was necessarily completed for a long time. It would not be in any way surprising if the Pearl cycle took the best part of the decade (or even more) to create. In the sphere of art, rather than of poetry, a similar such 'ultimate golden ratio' composition is *The Golden Stairs* painted by

Sir Edward Coley Burne-Jones in the nineteenth century. Containing close to 40 golden rectangles, it took nearly 10 years to finish.

241 Christine Fell (tr.), John Lucas (poems), *Egils Saga*, J.M. Dent & Sons Ltd., 1975

242 Alan Thacker, *The Cult of King Harold at Chester* (in *The Middle Ages in the North-West* ed. Tom Scott and Pat Starkey) Leopard's Head Press 1995

243 St. John's may be of some significance in identifying the author or at least the provenance of the story which inspired the plot of *St. Erkenwald*. In 1364, Jordan de Holme (who was a canon of St. John's) exchanged the parish of Stockport for that of Sefton held by John de Massey since 27 November 1339. Jordan de Holme is also said to have resigned at the same time from Ashton-on-Mersey in favour of a second John de Massey said to be from Sale and to have been ordained in June 1365. It would appear from the dates that the resignation of de Holme actually made a vacancy for the second John de Massey prior to his ordination. See also endnote 32 regarding the possible significance of the name Holme to that of the 'Green Knight'.

http://www.british-history.ac.uk/report.aspx?compid=41291 (accessed 02/08/08)

244 A 6:5 hexatonic division of the text occurs at Line 192. In the very next line, the corpse of the pagan judge speaks for the first time. A 6:5 hexatonic division of the first half of the poem occurs at Line 96 immediately before the first spoken dialogue of the poem—that of the workmen who find the corpse. This hexatonic musical division was very easy for the poet to create, given the already existing constraint that the text length was a multiple of 22—in fact, because of that constraint, it is much simpler than the musical twelfths, fifths and octaves of the *Pearl* manuscript. Another reason for believing that *St. Erkenwald* was a rush job.

245 Although this is not stated in the poem, it is almost certainly Barking Abbey which the historical Erkenwald had built for his sister in the year 666. However much 666 would be an interesting allusion in this poem, it seems unlikely it is there. The Line 112 could perhaps be considered as a reference to the year 1178, assuming that it is correct that the start of the poem alludes to the year 1066. 1178 is the year in which five monks of Canterbury testified on oath they saw a fire explode on the moon and, for no better reason, than the sheer joy of mentioning other meteorite falls, I will point out that Barking is noted for just such a celestial visitation in the year 664.

246 If this was indeed his intention, then his mathematics is remarkably prescient. The golden ratio and the fate–freewill conflict came to prominence in twentieth century business with a theory of stock market fluctuations based on an empirical economic model known as Elliott Wave Theory. A connection between the Fibonacci numbers (hence the golden ratio) and human decision-making is integral to this theory, which has long been realised to have radical implications for any scientific study of freewill.

247 Britton J. Harwood, *Gawain and the Gift*, PMLA, Vol. 106, No. 3. 1991

248 Indeed the parallel is quite explicit: mention of the fifth wound of Christ is found in Lines 649–660 of *Pearl*, a line numbering which in *Sir Gawain and the Green Knight* corresponds to a discussion of the fivefold fives of virtue after a mention of the Five Wounds of Christ in Line 642. See also in this regard Endnote 47 and 52.

249 In fact, this avoids the one and only multiple of 10 that I can find, other than 490, in the entire manuscript. This may, of course, be simple serendipity rather than a conscious avoidance of multiples of 10.

250 All this is addition to compensating for zeroes in *Pearl* which pairs *wyrd* from Line 249 with Line 271, which then creates a field of 17^2 needed to reach the end, rather than a number referencing a synodic year.

The weaving of the text is even closer as shown in the following table which indicates shows two ways to get to Line 273 where 'wyrd' (fate) is mentioned: one way is through mathematical metaphors for *trawthe* and *dulcarnoun* (the twin horns of a dilemma symbolised by the solutions to the famous Pythagorean theorem, $a^2 + b^2 = c^2$), the other through the arithmetic tokens for *resurrection* and *fortune*. The resurrection number alternatively can be used through the number for a *year* (albeit a synodic one) to reach the *end*.

Line	Link	to Line	Link	to Line
		↱	× 22 ÷ 7	1212
1010	× 0.382	386	× √2 ÷ 2	273
1011	−153	858	× 7 ÷ 22	273
		↳	+ 354	1212

251 The combination of letters, aleph–taw, has two major different interpretations in Hebrew grammar. There's a choice between a Jewish rabbinical understanding or a Christian linguistic one. In the latter instance, there is no significance to the

word beyond the definition of an untranslatable particle used to mark the definite object of the verb. For the rabbis, however, it is difficult to find a two-letter word of greater significance. Aleph-taw (or alef-tav) is frequently found at the end of lines in ancient scrolls: the two letters are considered 'witnesses' on behalf of the scribe that the text has been accurately copied. Alef-tav thus symbolises truth. Moreover, for the rabbis, since alef-tav is comprised of the first and last letter of the Hebrew alphabet, the combination also symbolises the alphabet as a whole and thus, in addition, words. Because alef-tav also occurs in Genesis 1:1 in the phrase *bereshit bara Elokim et hashamaim v'et haarets* before the mention of the first created thing, the heavens, its positioning indicates that words are the very 'stuff' of creation. Words and letters therefore not only precede the creation of all things, they also describe the method of creation as well as form the vessel of creation. They give shape to creation and are also witnesses to it. (David Patterson, *Hebrew Language and Jewish Thought*, Routledge 2005)

252 There is no indication that even Fibonacci was aware of this particular point despite many claims that he discovered the golden ratio in his sequence.

253 EI Condren, *The Numerical Universe of the Gawain-Pearl Poet: Beyond Phi*, University Press of Florida 2002 p. *ix*

254 17.6 inches has long been regarded (including in medieval times) as the length of a 'standard' cubit in the Hebrew measurement system. The 'royal' cubit was 2.8 inches longer.

255 There was a distinct theological advantage in changing from 354 to 352, as well, since as noted previously in the case of *Sir Orfeo*, an exact multiple of 22 or 11 was not strictly *de rigeur* when constructing a 'Wheel of Fate' design. Assuming that *St. Erkenwald* was indeed written for the completion of the 'New Work' at St. Paul's cathedral, there was a small, but nonetheless significant danger, that anyone penetrating the numerical design of the poem might think the poet was suggesting that London was the City of God and St. Paul's the heavenly temple. It is likely the poet would have recognised this possibility and it may have influenced the change. For all his anti-French pro-British sentiment, he does not have the extreme nationalistic fervour which characterises the patriotism of later centuries.

256 JRR Tolkien, *Sir Gawain and the Green Knight* in *The Monsters and the Critics and Other Essays*, Harper Collins 1983, p105

257 Within this same digression about the armour, at Line 632, we are told that

Gawain is trustworthy in five ways and fivefold in five ways, a knight as virtuous as pure gold. A double golden section being some five lines away, it is a momentarily perplexing to find this 'error' of positioning with respect to a mention of gold. Five lines is at least four too many, in my view. However, the small discrepancy can be accounted for, in my opinion, by noting that 1743 − 1111 = 632. By a shift of five lines, two references to the Kingdom of Heaven are implied. The digression as a whole is a knotwork of fettled mathematical metaphors.

However I would like to suggest we can take this same idea further. Assuming that *Pearl* is linked in the same way to *Sir Gawain and the Green Knight* as it is to *Patience*, then we should automatically expect that there would be a significant reference to the Four Daughters of God on either Line 531 or 632 (depending on whether *Pearl* is counted as 1111 lines or 1212) Line 531 says 'no deceit' – perhaps a reference to truth/trawthe. But frankly one out of only Four Daughters is not sufficient. On the other hand, Line 632 which we are considering here, may have, according to Andrew and Waldron, 'poynte' implied. If 'poynte', then zero: if zero, then a mathematical token for a kiss in the poet's ideological set–up. I tick this off as yet another fettle.

258 'One would expect a name that glorifies the speed and sailing qualities of the ship, instead of a name that...is apt for a shield with a female head painted upon it. Though it is possible that a legendary shield and a ship became confused in the oral transmission of Arthur's story... 'Pridwen' could be a distorted memory of an ancient Celtic war tactic. The members of the army of the Celtic leader Brennos used their long shields as rafts to aid them in swimming across a river in 280 BC (Ellis, P. B., *Celt and Greek*, London, 1997, p14). A memory of this tactic may have led to a shield, with a painted female face upon it, becoming confused with a ship. With the consequence that the name of the shield would have been transferred to the ship, leaving both with the name Prydwen.' (Michael Wild, dagonet_uk@yahoo.co.uk accessed 28/09/09)

259 Now while this is not the same Brennus as the brother of Belin, it's my view that the poet didn't know that. If he has an error in his scholarship, I think this is one of the very few. I believe he purposely named this obscure Celt precisely because Brennus was the leader of the tribes who sacked Delphi. Although we now know this happened around 279 B.C., the poet used the sources available to him to date it to 354 B.C., in the era of Brennus, brother of Belin, not realising there were two warlords from Gaul of the same name within a generation or so of each other, one who devastated Rome and the

other who devastated Greece.

260 Although Ross Arthur sees Gawain's shield as a Shield of Truth, it seems more probable to me that the poet envisaged it as the Shield of Faith, in alignment with the portrayal of a shield in the famous passage in Ephesians describing the Armour of God. This would suggest that the green girdle is meant to be perceived as a foul substitute for the Belt of Truth, thus making the Endless Knot or pentagram on the shield more properly analogous to Truth than the actual shield itself.

At the risk of adding confusion to this analysis, it should be noted that Gawain's shield also bears some resemblance to the shield of Aeneas, the Trojan exile, who is mentioned so prominently on Line 5 of the poem and who is claimed as Gawain's forbear. The description in Virgil's *Aeneid* shows that by Book VIII, armed by Venus with Vulcan's magical armour, including a shield depicting a pageant of Roman heroes, Aeneas is ready to take on his opponents. His arming reflects that of Achilles in Book XVIII of the *Iliad*.

Patrick Reid points out the contrast between classical heroism and faith in God as a covenant defender which may have some bearing on the poet's thought. (Patrick V Reid, *Moses's staff and Aeneas's shield: the way of the Torah versus classical heroism*, University Press of Arizona, p.10) From the beginning of *Patience* to the mention of Aeneas in the fifth line of *Sir Gawain and the Green Knight* is 536 lines. Divide this by the golden ratio to get 867—and 867 lines further along is, according to George Sanderlin, another reference to Aeneas in the transfiguration of Gawain after he arrives at Hautdesert. The passage allegedly reflects that in the *Aeneid* where Venus dissipates the mist to reveal Aeneas to Dido in the temple at Carthage. (Not too much can be read into the parallel use of the golden ratio in the *Aeneid*—as found by George Duckworth— and in *Sir Gawain and the Green Knight* as far as I am concerned, since this is such a common mathematical trope.) Again applying the golden ratio within *Sir Gawain and the Green Knight*, we move to Line 1408, which specifically mentions covenant. Was the poet thinking of Aeneas' tawdry behaviour towards Dido and his insistence that there was never a covenant between them? Unless the golden ratio linkage between the references to Aeneas in *Sir Gawain and the Green Knight* are entirely coincidental (which seems highly unlikely to me), then there is also some thought of what a covenant truly means mixed in here. That *trawthe* alone is needed for God to act as Gawain's covenant defender; magic will not serve. Perhaps there is some irony in Brennus' victory over

Rome when Aeneas' shield is a pageant of Roman heroes.

It's also worth noting another possible mathematical link here based around another more obscure allusion to Aeneas: he was a prince of Troy, a placename mentioned in the very first line of *Sir Gawain and the Green Knight*. Throughout the Middle Ages, 'Troytown' was the name given to various labyrinths cut in turf or outlined in stone across Britain and Northern Europe. They were also called Gillian–mazes or Gillian–, Gelyan–, Julian Bowers (Gilling Bore, Jul–Laber) possibly deriving from Ilium, the name given to Troy's inner keep or palace. (Theodore Silverstein, *Sir Gawain and the Green Knight*, A New Critical Edition, University of Chicago Press 1974, p.112) Another possibility for the origin of the name is given by Dr. Stukeley who suggested that the term Julian's Bower was derived from the name of Iulus, the son of Aeneas, who is described as having taken part in a game involving labyrinthine moves.

http://www.robertabarresi.com/mazeslabyrinthcap12.html, accessed 17 November 2011

While St Julian (Gilyan) is mentioned in *Sir Gawain and the Green Knight*, it is difficult to ascertain whether its placement is significant in terms of a labyrinth. It occurs on line 774 which is 13 × 333 lines into the text.

However, if we look for a 'Troytown labyrinth' using references to Troy as our mathematical base, then we should note that it is 531 lines from the beginning of *Patience* to the first reference to Troy in *Sir Gawain and the Green Knight*. This, when divided by the golden ratio, is 859 lines—one more than the number of feet in the labyrinth at Chartres Cathedral and one more than the significant Wheel of Fate inside *Pearl*. Line 859 of *Sir Gawain and the Green Knight* does refer to footsteps on the floor, suggestive of the possibility of walking a labyrinth. But only vaguely. It is here he loses the last of his pentangle devices and the use of 'disployed' may be an echo of the conquest of Troy in Line 6. This is the very section George Sanderlin believes refers to Aeneas. So: if there is a specific mathematical metaphor within this design, and if it does point to the Chartres labyrinth, then it is likely that the clue points to Gawain entering a monster's maze. In the Middle Ages, the minotaur was pictured at the centre of the Chartres Cathedral labyrinth.

It should also be noted that, in another Roman connection, Julius Caesar mentioned Chartres as one of the three alternating towns where the druids of Gaul held their annual assemblies. Like Troyes, it was a long–term centre of druidism and of neo–Platonism and Pythagoreanism.

Adding to the possibility that there is an elaborate labyrinth (which seems to be a logarithmic spiral) is the fact that the first reference to Aeneas is 536 lines into the second 'half' of the manuscript and this, multiplied by the golden ratio, is 331. At Line 331, Arthur has just picked up the Green Knight's axe and is about to strike a blow. Gawain is set to emerge as the hero of the poem as he stands and, with punctilious courtesy, steps in. The key element here is the mention of the axe in Arthur's hands. The word 'labyrinth' is considered to derive from *double-headed axe* and it seems very likely the poet is again playfully punning within his mathematical design. [This suggests that the poet knew the origin of the word, which in later times was obviously lost, until its re-discovery in the nineteenth century as evidenced here: 'Down to a few decades ago we were content with the bald statement of most dictionaries that it was probably correlated with the word laura, meaning a *passage*, or *mine*, though there was also a suggestion that it might be of Egyptian origin, viz., that it was derived from the name of Labaris (= Senusret III), erroneously conceived by the scribe Manetho to be the founder of the Hawara pile. Then Mr. Max Mayer put forward the suggestion that it might have some connection with *labrys*, a word which, in some of the early languages of Asia Minor, e.g., Lydia and Caria, denoted an axe, the axe being the symbol associated with the god known as Zeus Labrandeus or Zeus Stratios, the worship of whom was known to have taken place at Labranda, in Caria. Coins from Mylasa, a neighbouring town, show this god holding in his hand a double axe.' William Henry Matthews, *Mazes and Labyrinths: Their History and Development*, NABU Reprint of 1882 edition]

261 Perhaps it was this seeming lack of cooperation from the oracle that prompted Nero to appropriate hundreds of statues from the sanctuary at Delphi for himself. Or perhaps he simply fancied lots of the finest Greek marble.

262 The 'divine white' navelstone at Delphi was not the only stone which was considered to indicate the 'navel of the world'. Another oracle, as famed as that of Delphi, also laid claim to that title. 'Alexander expressed a wish to be buried at the Oracle of Siwa where a meteorite was the stone of prophecy. This one was beset by local emeralds, exactly as the green *lapis exilis* fallen from heaven would be represented as the Grail in Wolfram von Eschenbach's *Parzival*.' Andrew Sinclair, *The Discovery of the Grail*, Arrow Books, 1999, p.12 (Which leads me to wonder about the ship of Solomon in all those Grail stories: can 'ship' be a pun for 'shape' and thus for the Endless Knot?)

263 With Delphi and its worship of Apollo, the sun god, as a constant background thought, it's no wonder the *Pearl* poet wrote as if *Erec et Enide* were a marriage of the sun and moon, rather than a marriage of the moon and earth, when he clearly knew better.

264 *Purity*, Line 1525. Anderson considers the major divisions of *Purity/Cleanness* to be:

- The story of the Flood from Lines 249 – 544 (296 lines)

- Transition of 56 lines

- The story of Sodom and Gomorrah from Lines 601 – 1052 (452 lines)

- Transition of 104 lines

- The story of Bel's Feast: 1157 – 1804 (648 lines)

It should be noted that 153 (the 'number of the fish' and the seventeenth triangular number) is the probable determinant for the length of the Bel's Feast, since $153 \div \tau^3 = 648$ and also that $648 \times \tau^2 = 400$ which is the length of the story of the Flood plus the second transition.

265 John L. Lowes, *The Prologue to the Legend of Good Women as Related to the French Marguerite Poems, and the Filostrato*, PMLA, Vol. 19, No. 4, Modern Language Association, 1904

266 David Maw in *Machaut and the 'Critical' Phase of Medieval Polyphony* (*Music and Letters*, vol. 87, no. 2, Oxford University Press) pointed out the use by Machaut of 7 and 49, 12 and 144, the first two of which he suggests are numbers of forgiveness. The combination is reminiscent of the *Pearl* poet's formulation.

267 Notwithstanding, the suggestion of John Bowers that *Pearl* is a lament for 'good queen' Anne of Bohemia, wife of Richard II, is worth considering.

268 The golden hair in *Cligés* belonged to Sordamour, the sister of Gawain. In no other romance does this particular character appear. A lengthy exposition of the meaning of Sordamour's name reveals that it is connected with love (*amour*) and gold/golden hair (*sor*, meaning *bright blonde*) and that 'covered over with gold' would not be an inappropriate translation of her name.

269 Loomis argued that Welsh Gwalchmai, *Hawk of May* or *Hawk of the Plain* was originally a different character to Gawain and he was in fact Gwrvan Gwallt-avwy from the *Mabinogion* whose name may have arisen from gwallt-avwyn or gwallt-advwyn. Other scholars however consider Gawain and Gwalchmai

to be identical. In some versions Gawain's story has a marked resemblance to that of Pope Gregory: the illegitimate child of Arthur's sister, Morgause, and her page, Lot (later to be king of Orkney or Lothian), he was baptised and set adrift in a cask. Rescued by fishermen, he eventually found his way to Rome where he was knighted by Pope Sulpicius. In early romance he was not only the sister-son of Arthur (and thus his heir), but the pre-eminent champion of Camelot. Later stories, particularly the French ones—and Malory, who was influenced by them—depict him as a rapist and a bully. It is often noted that Gawain had a peculiar aspect: he grew stronger towards noon. This has led to him being considered as a solar hero or as a faint memory of a Celtic sun god. Jacob Jacobs, however, notes the similarity between many traditional tales of the hero Owain and those of Gawain.

The ballad *Kemp Owyne* (Owain the Champion) recalls the magic, the gifts of invulnerability and the testing kisses of *Sir Gawain and the Green Knight*. Found in Buchan's *Ballads of the North of Scotland* is one of many folkloric versions of this ballad which have suggested to collectors like Jacob Jacobs that Owain, the son of Urien Rheged, a sixth century king of Welsh-speaking Scotland is fundamentally the same mythic personage as Gawain, nephew of Arthur. Moreover the parallels in *Kemp Owyne* for the magic girdle of *Sir Gawain and the Green Knight* might have been, in other variants of the basic story, a ring, a belt or a sword. These versions include Icelandic, Danish, and German tales as well as the Northumbrian *The Laidley Worm of Spindleston Heugh* and *The Laily Worm and the Machrel of the Sea*. The ending of the *Kemp Owyne* is comparable to the ending of *The Wedding of Sir Gawain and Dame Ragnell* and at 62 lines hints at a golden ratio sub-structure:

KEMP OWYNE
(from Ballads of the North of Scotland)

Her mother died when she was young,
Which gave her cause to make great moan;
Her father married the warst woman
That ever lived in Christendom.

She serv-ed her with foot and hand,
In every thing that she could dee;
Till once in an unlucky time,

She threw her in ower Craigy's sea.
Says, "Lie you there, dove Isabel,
And all my sorrows lie with thee;
Till Kemp Owyne come ower the sea,
And borrow you with kisses three,
Let all the warld do what they will,
Oh! borrowed shall you never be."

Her breath grew strang, her hair grew lang,
And twisted thrice about the tree;
And all the people far and near,
Thought that a savage beast was she;
These news did come to Kemp Owyne,
Where he lived far beyond the sea.

He hasted him to Craigy's sea,
And on the savage beast looked he;
Her breath was strang, her hair was lang,
And twisted was about the tree;
And with a swing she came about,
"Come to Craigy's sea and kiss with me.

"Here is a royal belt," she cried,
"That I have found in the green sea;
And while your body it is on,
Drawn shall your blood never be;
But if you touch me tail or fin,
I vow my belt your death shall be."

He stepp–ed in, gave her a kiss,
The royal belt he brought him wi'
Her breath was strang, her hair was lang,
And twisted twice about the tree;
And with a swing she came about,
"Come to Craigy's sea and kiss with me.

"Here is a royal ring," she said,
"That I have found in the green sea;

And while your finger it is on,
Drawn shall your blood never be;
But if you touch me tail or fin,
I swear my ring your death shall be."

He stepp–ed in, gave her a kiss,
The royal ring he brought him wi';
Her breath was strang, her hair was lang,
And twisted ance about the tree;
And with a swing she came about,
"Come to Craigy's sea and kiss with me.

"Here is a royal brand," she said,
"That I have found in the green sea;
And while your body it is on,
Drawn shall your blood never be;
But if you touch me tail or fin,
I swear my brand your death shall be."

He stepp–ed in, gave her a kiss,
The royal brand he brought him wi';
Her breath was sweet, her hair grew short,
And twisted nane about the tree:
And smilingly she came about,
As fair a woman, as fair could be.

The magic ring, magic belt and magic sword are simply three variations of the same symbol and are paralleled by the green girdle in *Sir Gawain and the Green Knight* which had precisely the same property of rendering the wearer invulnerable. In another variant on a similar theme, the shape–shifting of the enchanted woman has become an attribute of the Green Knight. This, of course, raises a possibility as to why *Sir Gawain and the Green Knight* was never widely distributed if it was finished, as I suspect it was, around the year 1400.

'Gawain' is almost certainly, despite the difference in meaning given in modern books of names, a variant of 'Owain'. How could a West Country poet safely release a poem about the perfect knight, Gawain, during the rebellion of Owain Glendower? How would it have been seen? Even if the poet sympathised with Glendower, he might not have wanted it to be seen that he was intimating that

Glendower was without fault. The poet, unquestionably, has some distinctly Welsh leanings. The twentieth century's idealised dream of Camelot is reflected in the poet's vision of Logres, mentioned only on Line 690 of *Sir Gawain and the Green Knight*. This is 1221 lines from the start of the second 'half' of the manuscript. That is, it is 11 × 111 lines from Line 1 of *Patience*—and with that string-of-ones, it is evident the poet is using Arthur's kingdom as a mirror of the kingdom of heaven—but he is also using a Welsh name for it.

'Logres' might be mediated via the French but it was still Welsh in origin. The beginning of the fifteenth century was a extremely dangerous time to be pro-Wales. It may be suggested that, to salvage the poem, the name Gawain could simply have been changed. Anyone who thinks this way has obviously never written fiction. For many authors, finding the right name for a character is the most important preliminary. Some authors simply can't start unless they have the right names. I have met quite a number and have been reassured about my own issues in this regard in speaking to them: spending months in dictionaries of first names is not all that unusual. When I heard Michael Morpurgo—former children's laureate and author of *Warhorse*—say that it might sound pathetic to admit that he couldn't begin a story without the right names, I sympathised totally. True, it may sound like a feeble excuse—but that's the way it is for many fiction writers. Perhaps within some of us it's to do with a deep sense of the ancient belief that names are carriers of destiny, not interchangeable labels. I know that's my prime reason. So, I'm not convinced that the *Pearl* poet could have substituted another name here, just to overcome the political overtones of Gawain and it's relation to Owain. To suggest that human creativity has changed so much in six centuries is not credible. Moreover, while it may not have mattered too much in other circumstances, it did here because the poet's hero had to be the antithesis of Lancelot. To invent an all-new hero was probably unthinkable for him.

270 Robert R. Edwards, *Ratio and Invention: A Study of Medieval Lyric and Narrative*, Vanderbilt University Press 1989, p. 121

271 Chaucer's 'retraction'—his repudiation in *The Canterbury Tales* of portions of his earlier poetry as ultimately unchristian—is seen by many of his admirers as highly embarrassing but it may well be explained by deeply hidden neo-Pythagoreanism planted within the verses. Reverence for Delphi's navelstone rather than for Jerusalem's Cross may well have been a characteristic of his work. Whether he is the only major English poet who worked in this way must remain an open question. John Milton who, as mentioned previously, was

well aware of numerical literary technique seems to suggest in his sonnet on Shakespeare that the Bard's sympathies were so inclined:

> *Thy easy numbers flow, and that each heart*
> *Hath from the leaves of thy unvalued book*
> *Those Delphic lines with deep impression took*

'Thy easy numbers' and 'Delphic lines' suggest that Milton recognised a neo-Platonic Pythagorean sub-structure in some, if not all, of Shakespeare's work.

272 Indeed the structure of *Erec et Enide* is clearly built around both the golden ratio *as well as* the 1111 noted by Joan Helm. 'Ici fenist li premiers vers' Chrétien tells his readers on Line 1796: *here ends part one*. It is far too remarkable a coincidence that 1796 × τ = 1110, only 1 line different from the perfect golden section. (This tiny discrepancy also occurs in *Le Chevalier de la Charette*, despite Joan Helm's contention that it is divided in a perfect golden section. The mysterious decorated E which puzzled her occurs on Line 4401, which is almost, but not quite, the golden ratio of the total line count, 7118. The golden section is 4399.16 which suggests it is more appropriate to be on Line 4399 or since the mathematical pedants would note it goes 0.16 into the next line, Line 4400.)

273 There is a curious connection between Chrétien's use of 540 here and the Old English poem, *The Phoenix*. This link suggests that Chrétien was in fact reworking a much older format—one that Chaucer tapped into a few centuries later again. Like Chaucer's *The Parliament of Fowls*, *The Phoenix* has an unusual asymmetrical structure. As might be expected, given the traditional lifetime of the phoenix, this structure is based around the number 540. On Line 540 of *The Phoenix*, the bird is reborn. Despite this superb stichometric timing, the poet was up against an unusual problem in designing the poem. The very peculiar number of lines from Line 540 to the end of the poem is clearly dictated by a structural difficulty with theological overtones. There are 677 lines in the poem and thus 137 from Line 540 to the end. The problem lies in the fact that 540 × τ = 333, which from the poet's point of view was no doubt absolutely fabulous. I believe it evoked (as I have already suggested for Chaucer's *The Parliament of Fowls*) the Garden of Eden by being a multiple of 3 (for Trinity) and 111 (for City of God). Unfortunately, however, a symmetrical poem would dictate a line length of 2 × 333 or 666 lines—thus aligning the phoenix with the Beast and the Anti-Christ in the Book of Revelation. Not a happy consequence of working with symmetry! However, a bit of number juggling by the poet

314

would have produced an interesting and much more acceptable result: $^{137}/\tau =$ 222. The 137 lines which end the poem thus come from finding the *tau* of 222. However since these 137 lines are the finale, it is appropriate to measure the 222 back from the end, thus arriving at Line 455. This corresponds to 122 lines (or 11 × 11 + 1 lines) from the *tau* of the first part, which makes it a workable compromise in terms of keeping only to mathematical metaphors of 'heaven'. While we're here, some interesting mathematics to note:

$$540 = 77 \times 7 + 1$$

$$77 \times 7 \times \tau \approx 333$$

$$55 \times 5 \times \tau \approx 170$$

$$22 \times 2 \times \tau \approx 27 = 33$$

$$22 \times 2 \times \tau^2 \approx 17$$

$$17 \times 77 \times 2 = 2618$$

As regards the *Pearl* manuscript, the phoenix is mentioned in *Pearl* on Line 430, where the Virgin Mary is called the 'Fenyx of Araby'. (Even excluding or including zeroes doesn't shed any light on why it's Line 430 and not Line 540.) The placement seems strange initially and there is no obvious mathematical metaphor evident at first sight. However it is 5656 lines from the end of the manuscript and as this is a multiple of 101, it is in itself very suggestive. Still it fails the test of a metaphor because it is neither a multiple of 540 nor 54. Nonetheless it is worth looking in the vicinity for lines that are. The only one that qualifies is Line 2329 of *Sir Gawain and the Green Knight* which is 5454 lines (or 54 × 101) beyond the 'Fenyx of Araby' in *Pearl* and is the point where Gawain's ordeal abruptly ends. Thus the motif of Resurrection is linked back 5454 lines to the phoenix.

Moreover there seems to be a mathematical crux in Lines 2284–2287 of *Sir Gawain and the Green Knight*. I'm not entirely sure it points to a phoenix or a Resurrection motif, but there is a major clue that it is significant. In these lines several extremely significant words—*point*, *destiny* and *trawthe*—all cluster. 2287 is 243 lines (by exclusive counting) from the end and 243 is 3^5 which does feature the thematic fives of the poem. Also 2284 is $540 \div 0.618^3$ but whether either of these mathematical motifs was in the poet's mind seems far from clear. He is usually far more explicit in his intention.

Nonetheless I believe that something elusive of major significance is to be found here. What is is, however, I am unable to divine.

274 Christiane L. Joost-Gaugier, *Measuring Heaven: Pythagoras and His Influence on Thought and Art in Antiquity and the Middle Ages*, Cornell University Press, 2006

275 Clearly inspiring the stanza in Milton's *On Time*:

> *The Oracles are dumm,*
> *No voice or hideous humm*
> *Runs through the arched roof in words deceiving.*
> *Apollo from his shrine*
> *Can no more divine,*
> *With hollow shreik the steep of Delphos leaving.*
> *No nightly trance, or breathed spell,*
> *Inspire's the pale-ey'd Priest from the prophetic cell*

> Kenneth Gross, *Each Heav'nly Close: Mythologies and Metrics in Spenser and the Early Poetry of Milton*, PMLA, Vol98, No1, 1983

276 Given that the major extant work on the life of Pythagoras dates from some 800 years after his death and that it was given prominence precisely in order to undercut the Christianity of Constantine's empire, it is difficult to be certain of the provenance of any numbers used by both the Pythagorean Brotherhood and the early Christian community. What, for instance, are we to make of the fact that Paul's blockbuster 202-word sentence at the beginning of his letter to the Ephesians opens with a reference to God the creator and sustainer? Is 101 as a metaphor for 'creator and sustainer' a Hebrew concept? (Pythagoras, after all, has been accused of stealing his mathematical concepts from Moses—though personally I'd give more weight to the possibility he plagiarised from the prophet Daniel.) Or did Paul appropriate and re-define an already-existing Platonic concept associated with the Music of the Spheres? Or if he didn't re-define it, did he simply allude to it so that the Ephesians couldn't miss his intent? Any of these is a possibility and no definite conclusion can be drawn.

277 Rebecca Gaines, Holly Barbaccia, *Sir Gawain and the Green Knight*, Spark Publishing, 2002, p27

278 Claude Luttrell in *The Folktale Element in Sir Gawain and the Green Knight* points out the thematic relationship between the Beheading Game and the continuation to Chrétien's *Le Conte de Graal*, providing yet another strong link of inspiration (or counter-inspiration) between the two.

279 The other lines—more than 30 of them—appear to be randomly placed. This

would suggest that the 17, 51 and 68 lines are simply coincidental. However, as ever with the *Pearl* poet, there may be additional constraints in play which would clarify the choices, if they are indeed deliberate choices to exclude 'e'. This curious and persistent connection between 'e' and truth exists, not just in these Arthurian stories and in Greek worship of Apollo which Plutarch entwined with Platonism and Pythagorean mysticism, but also in Yiddish folklore. The famous tales of the rabbis who invented mechanical servants—golems—to do their bidding include a strange detail: the golem often had the Hebrew word for truth, *emeth*, on their tongue. To deactivate the golem, it was necessary to take the first 'e' from *emeth* to produce *meth*, the Hebrew word for death.

280 Whether this view of 17 goes back as far as Plato is difficult to ascertain. It is impossible to ascertain if it goes back to Pythagoras, since only fragments of contemporary accounts of his life and work are extant. The main sources regarding Pythagoras date from nearly eight hundred years after his death. Plato, however, does mention 17 in *Theaetetus* where he recounts that the geometer Theodorus examined the areas of squares to demonstrate both rationality and irrationality when it came to the length of the sides. In an ambiguous statement which has caused controversy from ancient to modern times, Plato reports that Theodorus broke off at a square with an area of 17.

281 Usually this is quoted as 16, but this ignores the word based on *kauchaomai* rather than *chara*, the former positioned (not surprisingly) to divide the latter into a 3:1 ratio.

282 There are 22 men, women or people mentioned. This shorter list of 17 excludes Cain and Esau who are not part of the roll–call, as well as excluding 2 generic instances of 'people' and 1 generic 'prophets'. With the exception of 'Moses' parents' the 17 are named. They include Sarah and Rahab, the positioning of which seem to be significant. Other uses of 17 in the Epistle to the Hebrews include 17 uses of εἰσέρχεσθαι, *to enter*, and 68 (4 × 17) of θεός, *God*.

283 In fact, this list simply mentions 17 instances of 17.

284 The Pythagorean concept of 17 as 'the barrier' seems to have influenced the counting numbers in both French and Italian. Most other languages have a change in counting style at 12, 20 or 60, but not French or Italian e.g. onze (11), douze (12), treize (13), quatorze (14), quinze (15), seize (16), dix–sept (10 + 7), dix–huit (10 + 8) etc

285 'Greek–speaking people in antiquity were users of the decimal system, hardly

in their way of writing numbers, but certainly in their way of pronouncing them'. M.J.J. Menken, *Numerical Literary Techniques in John – The Fourth Evangelist's Use of Numbers of Words and Syllables*, Supplements to Novum Testamentum, Vol LV, E.J. Brill, 1985 quoting Karl Menninger, *Zahlwort und Ziffer, Eine Kulturgeschichte der Zahl*, Göttingen, 1979

In fact, this way of speaking may have given rise to the speculation about a Greek origin of the ten 'Arabic numerals' which goes back to the 16th century in Europe. Nonetheless, there are many sources in Europe and the pre–Islamic Levant that attribute them to India. The earliest depiction of them in English, *The Crafte of Nombrynge* (c.1350), correctly identifies them as 'teen figurys of Inde'. (http://www.etymonline.com/zero.php)

286 Jesus said to them, 'Bring some of the fish you have just caught.' Simon Peter climbed aboard and dragged the net ashore. It was full of large fish, 153, but even with so many the net was not torn. Jesus said to them, 'Come and have breakfast.' None of the disciples dared ask him, 'Who are you?' They knew it was the Lord. Jesus came, took the bread and gave it to them, and did the same with the fish. This was now the third time Jesus appeared to his disciples after he was raised from the dead. (NIV)

287 From the time of Archimedes, the square root of three had been calculated from the intersection of two circles (a fish or almond shape) and approximated as $^{265}/_{153}$: whence 153 was known as 'the measure of the fish'.

288 The omphalos or navelstone at Delphi was covered by a marble casing which was decorated with a fillet or net. John may specifically have mentioned the unbroken net to allude to the navelstone at Delphi. While there can be no certainty of this, his purpose in writing the fourth gospel (so different to the others) was to specifically address Gnostic (including neo–Platonic and neo–Pythagorean) heresy, so it seems likely that deliberate, albeit subtle, comparisons would be there.

289 According to Plutarch (in *Pseudodoxia Epidemica*, Book III, chapter 25) who lived in the first century, Pythagoreans wouldn't eat fish. According to Porphory, Pythagoras once predicted the exact number of fish in a magnificent trawl and, on the contents of the net being counted, the fishermen agreed to do whatever Pythagoras said. Pythagoras released the fish back into the water, paid the fishermen and went his way. Porphory lived in the third century and had a vested interest in stories that ran counter to the descriptions of the actions of

Jesus in the gospels. The writer of both *Philosophy from Oracles* and *Against the* Christians, he was engaged in controversy with both a number of early Christians and also with his own student, Iamblichus, whose mystic theurgy was to go on to influence the emperor Julian. Writing also on astrology and musical theory, his most influential treatise has probably been *On the Impropriety of Killing Living Beings for Food.* It advocated for vegetarianism and against the consumption of animals and is cited with approval in vegetarian literature up to the present day. Porphory and Pythagoras were the two most famous vegetarians of the ancient world (though again, the views of Pythagoras cannot be ascertained with any degree of certainty.)

By the Middle Ages, however, their reputation as such was well-established. The vegetarianism of the Pythagoreans was not only famous throughout medieval times but was a matter of considerable condemnation during that period. It is therefore worth wondering, if only briefly, whether the minutiae of detail in the hunting scenes of *Sir Gawain and the Green Knight* is there for that reason—to continue the anti-Pythagorean theme signified by the continual use of 17.

290 In numerical literary style, every number appearing in the text itself is significant. So the comment that this is the '*third* appearance' of Jesus has to be taken into consideration. It may allude to the $\sqrt{3}$ or to the cubic function or even to both.

291 If he had, then 153 should be further divided into 125 and 28 (or alternatively, but less probably, 55 and 98), whereas it appears 153 is used principally in relationship to 231 or to the golden ratio, not in the measures suggested by Augustine. The number choices here are a more important issue than they seem. There is a tendency on the part of many scholars engaged in medieval studies to look unfailingly towards Augustine of Hippo as the source of an orthodox poet's theological interpretation. This could, in some cases— and this one in particular—be very unwise. Augustine found a compromise between Christianity and neo-Platonism. A millennium later, neo-Platonism had become a hydra with many heads. The *Pearl* poet could not find common ground with any part of the dragon and spent his time after cutting off one head simply whetting his axe to have a chop at the next. It is highly unlikely in these circumstances the *Pearl* poet would use any of Augustine's thought— but if he did, he would certainly avoid those parts of Augustine's commentary which reek of neo-Platonic numerology, as the interpretation of 153 does.

292 This technique of cubing and adding digits can, in particular circumstances, also produce 153 another way. Again it involves a three. When *most* 3–digit whole numbers which are divisible by three have the cube of their digits summed, then the cube of the resulting digits summed (and so on until the number does not change), 153 is produced.

293 666 would make a perfect mathematical token to symbolise the Pythagorean concept of re–incarnation on the following grounds: it has obviously got lots of sixes already, but it is also the 36th triangular number. 216 (= 6 × 6 × 6) was the pre–eminent 'circular number' of the Pythagoreans and thus symbolic of re–incarnation, but nonetheless 36 (=6 × 6) wasn't far behind. As a triangular number, it partook of the essence of the tetraktys and thus was a reflection of the Pythagorean music of the spheres and the ultimate truth represented by Apollo of Delphi. (It also linked back to the famous 'Theorem of Pythagoras' of which the simplest triad is 3, 4 and 5: $3^2 + 4^2 = 5^2$. If instead of squaring, the numbers are cubed, then the result is $3^3 + 4^3 + 5^3 = 6^3$ or 216.)

Pythagoras himself allegedly said that he was re–incarnated every 216 years, having in a previous life been Euphorbus, killed by Menelaus before the walls of Troy. Indeed, he is said to have recognised the shield of Euphorbus which had been dedicated by Menelaus to Hera. The followers of Pythagoras claimed a large number of re–incarnations for the sage, ranging from a great hero like Euphorbus to a lowly fisherman. Herein lies the essential difference in understanding between Christianity and Gnostic Pythagoreanism and why they were so deeply and radically opposed to each other. Re–incarnation involved the return of the personality to a different body; resurrection involved the return of the person to the same (but ideal) genetic structure.

294 Matthew 12:38–45

295 *http://www.newadvent.org/cathen/15714a.htm*

296 Since the dourad occurs *on* the appropriate line (using the rational approximation $^{22}/_{7}$) this suggests the poet actually did have an awareness of the irrational nature of the golden section and that it is in fact, more correct to place it on the next line.

297 It is also midway between two numbers associated with Pythagoras—667 in Chaucer's *Book of the Duchess* (and possibly in *Sir Gawain and the Green Knight*) as well as 618 as the traditional digits of the golden ratio. Furthermore it is 11 lines past the mention of Gawain as a man virtuous as pure gold on Line 632,

noted previously as being 1111 less than 1743. The knotwork in this particular section in fettling various parts of the manuscript together is not complex but it is very intricate.

298 Joost Smit Sibinga, *The Composition of 1 Cor. 9 and Its Context*, Novum Testamentum, Vol. 40, Fasc. 2, BRILL 1998

299 *http://www.british-history.ac.uk/report.aspx?compid=41291* (*accessed 02/08/08*) This is the same year Ralph Holmes (Sir Raoul Helme, once known as the 'Green Squire' and presumed to have, on his knighthood, been known as the 'Green Knight') lost his head defending King Pedro of Spain.

300 Paul's allusions to Delphi and indeed to the proposition that the E on the navelstone means 'if' are never more in evidence than in the famous 'love chapter': 1 Corinthians 13.

If I speak in the tongues of men or of angels, but do not have love, I am only a resounding gong or a clanging cymbal. If I have the gift of prophecy and can fathom all mysteries and all knowledge, and if I have a faith that can move mountains, but do not have love, I am nothing. If I give all I possess to the poor and give over my body to hardship that I may boast, but do not have love, I gain nothing.

Love is patient, love is kind. It does not envy, it does not boast, it is not proud. It does not dishonour others, it is not self-seeking, it is not easily angered, it keeps no record of wrongs. Love does not delight in evil but rejoices with the truth. It always protects, always trusts, always hopes, always perseveres.

Love never fails. But where there are prophecies, they will cease; where there are tongues, they will be stilled; where there is knowledge, it will pass away. For we know in part and we prophesy in part, but when completeness comes, what is in part disappears. When I was a child, I talked like a child, I thought like a child, I reasoned like a child. When I became a man, I put the ways of childhood behind me. For now we see only a reflection as in a mirror; then we shall see face to face. Now I know in part; then I shall know fully, even as I am fully known.

And now these three remain: faith, hope and love. But the greatest of these is love.

'If' is repeatedly used and the references to tongues of men and of angels, babbling prophecies, mysteries and knowledge, clanging gongs and mirrors are so apt a description of the cult surrounding the Apollo Python and the sibyl's prophetic utterances there can be little doubt as to Paul's target in these verses. The Pythia sat on a tripod over a crevasse where fumes (possibly hallucinogenic) drifted up. Her oracular words were taken down by the priests of Delphi who were then paid by the inquirer for their translation.

301 Such manicules were common from the twelfth to the eighteenth century. http://archbook.ischool.utoronto.ca/archbook/manicules.php (accessed 26/04/13)

302 Fistmeile or fistmele is a term meaning *fist measure*, which was used in medieval archery (and is still current today) as a method of measuring brace height. Fistmele is the breadth of the fist with the thumb stuck out, used to set the distance from the bow handle to the bowstring.

303 Perhaps, given this is a golden ratio cut, there is some thought similar to that of Kaitlin Fincher: *'what makes a fist well–built is the unity and harmony of different size and shaped fingers.'* http://prezi.com/fah2wmpycbbf/the–golden–ratio–project/ (accessed 26/04/13)

304 As noted previously, Chrétien tells his readers on Line 1796 of *Erec et Enide*: 'Ici fenist li premiers vers,' *here ends part one*. At this point, the hero has won his bride and announced the adventure of the 'Joie de la Cort'. Line 1796 seems to indicate a structure based on 1111 since $1796 \times \tau = 1110$. This again points back to Hebrew cosmology and further vindicates Joan Helm's analysis.

The *Pearl* poet does not appear to use 1796 as a structural number anywhere, but it is still possible he may be obliquely referring to Chrétien's manuscript since $1111 \div \tau^2$ may be one of the constraints on the second 'half' of the manuscript. There appears to be nothing 153 lines from the end of *Sir Gawain and the Green Knight* to confirm this conjecture, however. On the other hand, to confuse the matter, 153 lines from the beginning of *Pearl* (where the mathematics does not work nearly as neatly, only roughly) it might be possible to suggest that the 'first part' of the *Pearl* adventure has ended with the reaching of the impassable stream. The jeweller/dreamer is about to encounter a bride and, while 'Joie de la Cort' is not mentioned, it is perhaps alluded to in the 'wynne,' *joy*, of Line 154.

305 Translated by Malcolm Andrew and Ronald Waldron, *The Poems of the Pearl Manuscript – Pearl, Cleanness, Patience, Sir Gawain and the Green Knight*, Edward Arnold (Publishers) Ltd, 1978

306 Colin Spencer, *The Heretic's Feast: A History of Vegetarianism*, UPNE, 1995

307 Mador *of the Door* (or, *of the Gate*) accuses Guinevere of poisoning his cousin,

Patrise. As a consequence, he battles with Lancelot (who appears in disguise to defend the Queen's honour). This story occurs in the *Vulgate Mort Artu,* a part of the *Prose Lancelot,* a French Arthurian romance, dated as probably early thirteenth century. It also occurs in the English Stanzaic *Morte Arthur,* which may have existed in the fourteenth century. The only manuscript source for the *Morte Arthure* is the Lincoln Thornton Manuscript written in the fifteenth century by Robert Thornton, who copied an older text (now lost, which presumably derived from south-west Lincolnshire). Mador's story also appears in Malory's *Le Morte d'Arthur,* but this is too late to be the source of the *Pearl* poet's inspiration.

308 Condren points out that the poem has five dialogues and four monologues and that the first two dialogues add to 89 lines while the last three come to a total of 55 lines, making 144 altogether. 55 and 89 are well-recognised Fibonacci numbers which demonstrate an approximation to the golden ratio, clearly indicating the relative sizes of the monologues and dialogues was influenced by this factor. (Edward I. Condren, *Chaucer from Prentice to Poet: The Metaphor of Love in Dream Visions and Troilus and Criseyde,* University Press of Florida, 2008) Condren does not, however, note that the dialogues plus the monologues together total 793 lines. The *tau* of 793 is 490. This suggests that a plea for forgiveness is coded into the poem and that the *Pearl* poet's use of 490 as his defining number was not a startling innovation. Perhaps the question is: who was Chaucer seeking forgiveness for or from?

309 Jill Mann in *Price and Value in Sir Gawain and the Green Knight* suggests that the concept of price, prize, praise, value and exchange are all linked to a sense of proportion in *Sir Gawain and the Green Knight.*

310 The fact that the Green Knight rides in 666 lines from the beginning of the second 'half' of the manuscript might also suggest some deeper substance, linking the character to some aspect of neo-Pythagoreanism.

311 The *paz* of *Patience* is undoubtedly the most difficult 'non-event' to explain in the entire manuscript. Even to find what it's fettled to is a difficulty. Sure it's 202 lines into *Patience,* another multiple of 101, but there is nothing of major significance going on. The only suggestion I can make, and it's not really a solid one (because I can't discern the metaphor behind it), is that between the *paz* of *Patience* and the *paz* of the second half of the manuscript should be 965 lines—however, the poet seems to have, for no discernible reason, moved the *paz* (so that there is a

reference to the Golden Mass in *Sir Gawain and the Green Knight*) by either 4 or 5 lines. It has to be to make a 'fettle' and perhaps these two instances of *paz* are the link in question. 969 lines difference, rather than 965, is very significant, because 969 is the sum of the first 17 triangular numbers.

1 + 3 + 6 + 10 + 15 + 21 + 28 + 36 + 45 + 55 + 66 + 78 + 91 + 105 + 120 + 136 + 153 = 969

The sum of triangular numbers makes tetrahedral numbers (the triangles stacked on each other form a triangular–based pyramid or tetrahedron), so this would suggest a regular tetrahedron is encoded in the text. If Condren is right in suggesting that a dodecahedron is encoded in *Pearl*, it's likely there's a cube (perhaps accounted for by the 4913 or 17 × 17 × 17 found earlier) and octahedron lurking around as well. The poet, as an anti-Platonist, would not have overlooked the Platonic solids in his re–definition of the nature of the *kosmos*.

312 Jay Schleusener, *"Patience," Lines 35–40*, Modern Philology, Vol. 67, No. 1., 1969

313 Translated by Malcolm Andrew and Ronald Waldron, *The Poems of the Pearl Manuscript – Pearl, Cleanness, Patience, Sir Gawain and the Green Knight*, Edward Arnold (Publishers) Ltd, 1978

314 It seems likely to me that the first person to notice this in recent times was JRR Tolkien, although I can discover no place where he wrote about it explicitly. Yet it is clearly implicit in his description of the 'star–shaped' isle of Númenor which is explicitly described as a pentangle. In *Unfinished Tales*, Ar-Pharazôn, *the golden*, was the last king to rule Númenor. The inspiration for the name, Númenor, is unclear but it is perhaps a pun on 'gnomen' (both a mathematical shape and the upright of a sundial) and 'numinous'. The meanings of *pharez* and *phares* both cluster in the life of Ar-Pharazôn who was responsible for a defilement like that of Belshazzar which resulted not only in the loss of his kingdom but in the sundering of Middle Earth from the Undying Lands. Mene also appears in various forms, most notably in the name of the royal city, Armenelos.

Tolkien's friend and colleague, CS Lewis, on the other hand, seems to have noticed the opposition between Delphi and *Sir Gawain and the Green Knight*. Again, there is no record of his thought, only the inferences that can be drawn from his writing. His book, *The Silver Chair*, has been seen as drawing its inspiration from *Sir Gawain and the Green Knight* on several levels. It's not far from Lady of the Green Kirtle to Lady of the Green Girdle, after all! However, there are a number of parallels so far unrecorded: the Pythia–like nature of the Green Witch, with her dangerously ambiguous advice (even before she turns into a snake) is evident, but perhaps the most delicate touch is that the way to the underground realm is through the letter E.

315 Other Hebrew words for number are 'cheqer' (Job 26:26; Job 38:16) which has the sense *number/search* and the impeccably appropriate 'math' (Psalm 105:12) which has the sense *number/few*. The concept of division in some words became linked to the concept of sin; possibly because sin cuts the individual off from the community. Thus chet, *sin, guilty for sin*; chets, *arrow*, from chats, *arrow, divide, cut off*, from chatsah, *divide, cut in two* (as in chetsiy, *half* of Daniel 12:7); chetsrowniy, *enclosed* or *division of a song*; chata, *to sin, to miss, to miss the mark, to purify from sin*; chat, *divide, enclosure, cut, gather*.

316 Before 666 became so infamous as the Number of the Beast, 616 was held to be the dread mark of the devil. The change from one to the other was clearly very early because comments by Irenaeus indicate that even by the second century it was considered an error to hold to 616, not 666. The change was probably made in my opinion simply because, as already suggested, 616 was so easily swamped in the halo of 618, given that some numerological systems ignored differences of 1 or 2 between numbers. Hebrew thinking could encompass the idea of 616 as 'missing the mark' of 618 and falling short of perfection (see footnote immediately above), but Greek thinking may have trivialised the difference and confused 'truth' and 'sin'. 666 on the other hand has all those sixes and, moreover, is the triangular number of 36. The emphasis on 6 is so great that, coupled with the circular number 36 and the triangular number, it would have been relatively simple at the time to identify 666 as a warning about Pythagorean Gnosticism. 6 was, after all, their preferred number, particularly as it pertained to 6 × 6 × 6 or 216 as the number of reincarnation.

Curiously, the ultimate monster story—*Beowulf*—seems to be constructed using 0.616 to divide the poem, not in the usual fashion of the scenes involving Grendel and its mother to the scenes involving the dragon, but rather to almost perfect geographical division: scenes in Heorot to scenes in Geatland. Moreover, within this geographical division, 0.616 repeatedly marks the position of references to the ancient giants, the eotens. If this is an intentional arithmetic metaphor for Grendel and/or the dragon as a 'type' of the Beast of Revelation, it's a curious item of knowledge for a Dark Age writer to have had stored up in his wordhoard.

317 While we're with the Celts, it should be noted that there is some significance to the number 17—though, this pertains more to the Irish than the Welsh. The family of the Tuatha de Danaan (the Irish Gods) were, according to Caitlin Matthews, numbered in seventeen triads. 'The seventeenth generation was considered to mark the limits of ancestral which, reckoned three–four generations a century, puts at ancestral memory at between four and five hundred years.' (*The Celtic Book of Days*, Godsfield Press Ltd, 1995 p23) While this may impinge slightly on the poet's thinking, the wealth of Christian connotations of seventeen within the overall pattern of numbers indicates that its Celtic background would have been, at best, a peripheral consideration.

318 Is there even more to this than meets the eye? Is the poet suggesting that the discovery of the golden ratio rightly belongs, not to Pythagoras, but to the prophet Daniel? Such an assertion may not be as wild as it seems. Most of what is now known about Pythagoras dates from the fourth and fifth centuries A.D., some eight centuries after he lived. Roger Herz–Fischler has found no evidence of knowledge of the golden ratio by the Pythagorean Brotherhood until around this time; indeed the much–vaunted symbol of the Brotherhood, the pentagram, is not known before Lucian mentions it in the second century but even there it is not spoken of in relation to the golden ratio, merely as a symbol of health. So could Daniel conceivably have precedence over Pythagoras? According to the Book of Daniel, he was twice given authority over all the Chaldean magicians, soothsayers, sages, astrologers and dream–interpreters in Babylon: once in the time of Nebuchadrezzar and once in the time of Belshazzar. (See for example, Daniel 5:11 NKJV ~ *There is a man in your kingdom in whom is the Spirit of the Holy God. And in the days of your father, light and understanding and wisdom, like the wisdom of the gods, were found in him; and King Nebuchadnezzar your father—your father the king—made him chief of the magicians, astrologers, Chaldeans, and soothsayers.*) This position and authority would have automatically made him the chief of the magi. Now it so happens that it is from the magi of Babylon that Pythagoras allegedly learned of the irrational nature of the golden ratio—and this is quite likely, since the Babylonians were actually able to solve quadratic equations. They were aware that some quadratics did not have rational (whole number or fractional) solutions. It is inevitable they would have encountered it since the golden ratio is part of the solution of one of the most simplest–looking quadratics of all: $x^2 = x + 1$

Now, one of the few facts which is definitely known about the life of Pythagoras is that he was studying in Egypt when the armies of Cambyses invaded in 525 BC. He was one of the captives taken to Babylon. It was only sixteen years previously in 539 BC that Daniel had, by his own admission, interpreted the writing on the wall on the night before Babylon fell to the armies of Cyrus.

So, is there any evidence in the Book of Daniel for the golden ratio? In my view, it is difficult to say for sure. It depends on what the original meaning of 'phares' or 'peres' is. Although translated *division*, it is often used as the word for *half*. However, if it originally meant 0.6 instead of 0.5 (and this is not only far from impossible—in fact, given that the Babylonian emphasis was on sixes, tens and sixties, it is even extremely likely) then there are several instances of a number approaching very close to the digits of the golden ratio appearing in the Book of Daniel. These instances actually include *mene, mene, tekel, peres*, which are sometimes considered to hold monetary values. Changing *peres* to 0.6, these values are respectively 1, 1, $^1/_{60}$, $^6/_{10}$ which together total 2.61666666—less than 0.1% error from the accepted value 2.618 for the square of $1/\tau^2$. Moreover, it seems possible that Daniel was aware of the discrepancy, since the sum of the last two numbers he unveils is 2625 which, multiplied by this possible Babylonian value of τ, (0.616666), curiously yields 1618—the first four digits of modern *phi*.

319 Aaron E. Wright, *Gold and Grace in Hartmann's Gregorius* in *Medieval Numerology – A Book of Essays*, Robert L. Surles (ed.), Garland Publishing 1993

320 Take for instance the mention of Sir Bertilak and his hundred hounds in Line 1596 of *Sir Gawain and the Green Knight*. Previously cited in the same section as 40 dogs, such a change automatically suggests that this is a mathematical clue: indeed it is Line 5151 of the whole manuscript—not just a multiple of 101 (reflective of 1 man and 100 hounds), but the 101st triangular number as well as a multiple of 303 (the *tau* of 490) and 17. It hints at (or confirms) the important construction numbers of the text. This line has a significance that is not readily apparent since another construction seems to hinge on it. 554 lines into *Purity*, just 2 lines before the dourad of the manuscript (that excludes *Patience*) is a mention of beryl. Usually beryl, as a symbol of the Five Wounds on the Cross, is a clue word to a golden ratio calculation (as, for example, at Line 1011 in *Pearl* discussed previously) but the only position that the clue beryl seems to point to is the very close vicinity of Line 1596 in *Sir Gawain and the Green Knight*. It is not clear why this should be so, but it suggests that there

is more here than meets the eye.

Other mathematical features, either deliberately intended or serendipitously emerging from the overall design:

1. 1540 is the 55th triangular number. Line 1540 of Sir Gawain and the Green Knight has Gawain protesting about true love: not only is 1540 a triply triangular number (since 55 is also a triangular number, the 10th, while 10 is also a triangular number, the 4th) but it is a multiple of 22. Its dourad is none other than Line 490. Its paz is Line 589, the last line of stanza 25, where Gawain, having been partially and gloriously armoured, is about to head off to hear Mass.

2. Pythagoras' Theorem with its 'dulcarnoun' is famously used as the design of Chaucer's *Troilus and Crisedye*. Does it also apply to *Sir Gawain and the Green Knight*? If there is a dilemma, mathematically its equal 'horns' occur on Line 1790. This is immediately prior to Lady Bertilak's change of tactics when she begins to offer him gifts—culminating, of course, in the offer of the green girdle. However, the mathematics here is comparatively simple: it shows nothing of the diabolically clever way Chaucer uses 'dulcarnoun'—using the ratio of an irrational to a rational number (see El Condren, *Chaucer from Prentice to Poet: The Metaphor of Love in Dream Visions and Troilus and Criseyde* University Press of Florida, 2008.) Chaucer's virtuoso performance with the Theorem of Pythagoras does not appear to have any parallel in the work of the *Pearl* poet. Chaucer's mathematics at the point where he mentions dulcarnoun is excessively daunting; since so few people could solve the mathematical riddle, the blatant clue seems almost like a challenge. Or perhaps it's a boast.

    The *Pearl* poet on the other hand does not seem to set up any challenge of a similar nature. He uses elementary formulations, even when he is manipulating irrational numbers. He's not trying to hide his set of linked and locked relationships. While his work seems absolutely effortless by comparison with Chaucer's, the planning involved in order to achieve control of the locking and interlocking in the seamless web he has created is immense.

3. Another source of interesting mathematics is the superfluous lines in each poem, the 'remainder' over the nearest 100. In *Pearl* and *Purity*, there are 12 lines and in *Patience* and *Sir Gawain and the Green Knight*, there are 31 lines. The first 'half' of the manuscript has 24 lines above

3000 and the second 'half' has 62 lines. 24 is the *paz* of 62. The *tau* of 62 is 38: highlighted in *Purity* Line 1444 (or 38^2) by the mention of a crown (sometimes seen as a cincture) of gold.

4. 62 itself is the *tau* of 100. The 'remainder' or 'tag–end' of *Patience* (531 lines) and *Sir Gawain and the Green Knight* (2531 lines) over an even hundred is 31 lines. 31^2 is 961 and the *paz* of 961 in *Pearl* (which has a decorated initial regarded as a scribal error) is Line 368 which mentions 'endorde,' *gold–adorned*. Perhaps this unusual word 'endorde' is to clue any reader whose bent is to solve mathematical puzzles that this requires more work than usual: a golden section, a square root and a remainder would be involved in any solution starting from Line 368's 'endorde'. The decorated initial on Line 961 of *Pearl* is pointed out as a scribal error by Andrew and Waldron who consider the initial should have occurred on Line 973 at the start of the next section. They remark that the scribe presumably mistook Line 961 for the start of the new section because of the extra stanza in Section XV. 'Scribal error' is all too often invoked when in fact 'authorial clue' to the mathematical sub–structure would better fit the vast majority of cases. As previously noted, this decorated initial is at the beginning of Stanza 81, a significant number in the mathematical construction of a 'mystical union' of heaven and earth.

5. 153 as 'the measure of the fish' comes from the classical approximation for $\sqrt{3}$. When the entire 6086 lines of the manuscript is divided by $\sqrt{3}$, the 'excess' remaining beyond the first two poems of the manuscript is 490 lines into *Patience*, suggesting that it may be correct to see *Patience* as the sum of 490 + 40 + 1.

6. $1/\tau^{17} - \tau^{17}$ is exactly 3571 which, curiously, is also $3 \times 7 \times 10 \times 17 + 1$. In the manuscript of 5555 lines (where *Patience* is considered as zero,) Line 5355 (or 153×35) is 200 lines from the end of *Sir Gawain and the Green Knight* and corresponds to Gawain's cry of 'Stop!' (which, like Nature's cry of 'Peace!' in *Parliament of Fowls*, marks a transition to another movement of the poem.) The fact that in *Sir Gawain and the Green Knight*, Gawain's cry ends the Beheading Game on a line which is a multiple of 153 re–inforces the resurrection aspect of the overall theme.

7. $5555 \times \tau^5 \approx 500$ and *Patience* begins on Line 3025 of the manuscript which is 55^2

8. 1212 – 365 = 847. (365 being the number of days in an ordinary year, rather than a synodic year). If the zeroes in *Pearl* to that point are then removed, Line 847 becomes Line 777.

321 Or how much Chaucer was influenced by him. Coolidge Otis Chapman suggested that the similarities between *The Squire's Tale* and the first fitt of *Sir Gawain and the Green Knight* are not simply chance and that Chaucer was influenced heavily by his acquaintance with the former.

322 Recall the lines from Milton's ode *On Shakespeare* mentioned previously:

> *Thy easy numbers flow, and that each heart*
> *Hath from the leaves of thy unvalued book*
> *Those Delphic lines with deep impression took*

323 The author offers a clue that 17 is important to this poem in Line 41 where the dreamer climbs a mountain and is able to see 17 miles in every direction. This is clearly metaphorical, rather than actual, since in reality, the 'mountain' only needs to be about 56 metres high to achieve this vista.

324 *The Cambridge History of English and American Literature in 18 Volumes (1907–21) Volume II The End of the Middle Ages* http://www.bartleby.com/212/0135.html (accessed 01/09/08)

325 Sir John Stanley was appointed hereditary King of the Isle of Man in 1405 or 1406 and was also a member of the Order of the Garter c. 1405. It is his membership of the Order of the Garter, with its motto *Hony Soyt Qui mal Pence* (the last line of *Sir Gawain and the Green Knight*,) that has recommended Stanley to various scholars as the author. However, since he would have had to have written *St. Erkenwald* (assuming it was indeed written in 1386) while still in diapers, I discount him as a serious candidate.

326 aka: de Baguley

327 *http://bigelowsociety.com/rod/baghist1.htm* (accessed 01/09/08)

328 Louise M. Bishop, *Words, Stones and Herbs – The Healing Word in Medieval and Early Modern England*, Syracuse University Press 2007

329 Gawain himself seems unable to make this transition. He still clings to the letter of the law and is unable to absolve himself. (Rebecca Gaines, Holly Barbaccia, *Sir Gawain and the Green Knight*, Spark Publishing, 2002, p15)

330 Bates in *Reading and Believing: Covenant in the Poems of the Pearl Manuscript*

also points out the nick on the neck that Bertilak delivers to Gawain is also a cut reminiscent of covenant. It is very much like of the 'oth berith, the sign of the covenant God gave to Abraham: circumcision. The fact that it is delivered on the Feast of the Christ's Circumcision reinforces this idea.

331 1 Samuel 24: 1–12

332 Perhaps based on Isaiah 14:12–13. (Vulgate: *quomodo cecidisti de caelo 2lucifer qui mane oriebaris corruisti in terram qui vulnerabas gentes qui dicebas in corde tuo in caelum conscendam super astra Dei exaltabo solium meum sedebo in monte testamenti in lateribus aquilonis.* King James Version: *How art thou fallen from heaven, O Lucifer, son of the morning! How art thou cut down to the ground, which didst weaken the nations! For thou hast said in thine heart, I will ascend into heaven, I will exalt my throne above the stars of God. I will dit also upon the mount of the congregation, in the sides of the north.*) However, 'north' is also often associated in Scripture with the dwelling place of God. As for instance in Psalm 48:2 KJV: *Beautiful for situation, the joy of the whole earth, is Mount Zion, on the sides of the north, the city of the great King.*

333 Baillie points out that one of the reasons scientists have dismissed this interpretation of the cause of the Black Death is that, until the early twenty-first century, everyone 'knew' there was no such thing as a black comet.

334 A plain vanilla reading of *Hanes Taliesin*, unaffected by the sort of folding, spindling and mutilating of the text undertaken by Robert Graves in his influential *The White Goddess*, suggests that Taliesin is the name of a comet whose 'original country is the region of the summer stars'. The writer of *Hanes Taliesin* would appear to be citing the times and appearances of a particular comet by reference, not to years, but to significant events associated with its regular advent. It should be noted that, despite the modern belief that it was only during the Enlightenment that anyone realised that comets returned on a regular basis, much of *Hanes Taliesin* may well refer to the most famous of all hairy stars—Halley's Comet. For a long time, Halley's Comet re–appeared at 77.0 year intervals on average. Eight orbits thus would have equaled 616 years and the significance of 8 in this context is that the planet Venus would have been in precisely the same position when the comet re–appeared.

335 Baillie also suggests in *New Light on the Black Death* that perhaps the people of the fourteenth century were right in considering that the Black Plague and the black comet were causally linked.

336 John Speirs, *Medieval English Poetry—The Non–Chaucerian Tradition*, Domville–Fife Press, 2007

337 On June 18 or 19, 1178, five monks at Canterbury in southern England reported seeing a flaming torch spring up between the horns of the moon:

After sunset when the moon had first become visible a marvellous phenomenon was witnessed by five or more men...now there was a bright new moon...its horns were tilted towards the east; and suddenly the upper horn split in two. From the midpoint of the division a flaming torch sprang up, spewing out, over a considerable distance, fire, hot coals, and sparks. Meanwhile the body of the moon which was below writhed, as it were, in anxiety ... the moon throbbed like a wounded snake. Afterwards it resumed its proper state.

Isaac Newton was aware of this testimony from the medieval chronicle of Gervase of Canterbury, and in 1976, the geologist Jack Hartung was responsible for drawing attention to Newton's 're–tweeting' of this record. Hartung went on to make the remarkable suggestion that the monks had actually witnessed the formation of what is widely considered to the youngest lunar crater, Giordano Bruno.

Hartung's theory was quite popular for a couple of decades, despite some criticism from Nininger and Huss who suggested that perhaps the monks had been mistaken—that their 'lunar impact' was an exploding meteor high in Earth's atmosphere which was coincidentally in the observers' line of sight of the moon. Paul Withers of the University of Arizona agreed with them that Hartung had been far too hasty, and in 2001, pointed out that, since Giordano Bruno is actually 22 kilometres across, anything of that size should have produced ejecta which, when flung into space, would have resulted in a meteor storm on Earth which lasted several weeks.

No less august a body than NASA finally issued a statement that no lunar impact happened in June 1178 after Withers carefully checked all around the world and discovered that 'historical records show nothing, including the European, Chinese, Arabic, Japanese and Korean astronomical archives.' And there the scientific consensus rests.

Now far be it from me to quibble with such eminent researchers but, well, I have noticed something that I would have thought NASA should have realised long before I did: *the world doesn't actually stop at the Equator.* Those historical records seem to be a just a little biased in their interpretation of what constitutes 'the world'. I should therefore like to point out that, preserved in the oral history of the Māori of New Zealand, is this extraordinary fragment dating back approximately eight hundred years:

> *Tahuri mai o mate te tihi ki 'Tirau*
> *Mowai rokiroki, ko te huna i te moa;*
> *I makere iho ai te tara o te marama e–i.*

> *Turn your eyes to the peak of 'Tirau*
> *Now desolate, utterly; all is lost,*
> *Gone, vanished as the moa;*
> *Oh, broken is the tip of the crescent moon.*

The translation of this lament is taken from Michael King's *Te Ao Hurihuri*. As Keri Hulme notes in her short story collection, *Te Kaihau – The Windeater*, where she plays multiple word games with Māori 'tara' (here translated *tip*), it actually has 21 different meanings. One of them is *horn*. So if we substitute *horn* for *tip* in the last line of this lament, it refers to a breaking of the 'horn of the moon'. The lament is traditionally used in late June, the same time as the five monks were said to have seen the writhing smoke on the horn of the moon. It also refers to the extinction of the moa (a large flightless bird whose demise the Māori themselves attribute to a strange wildfire from heaven known as the 'Mystic Fires of Tamatea', although anthropologists of course attribute it to over–hunting). Wildfire from heaven linked to a 'breaking of the horn of the moon': phrasing it like this makes the lament virtually identical in both specific and general description in the chronicle of Gervase of Canterbury.

I should also point out that, even in the northern hemisphere, the records aren't as totally devoid of celestial apparitions as Paul Withers suggests. The Irish *Annals of the Four Masters* relate: *Kaphoe died in the year 1178; it is stated that 'a great miracle was performed on the night of his death, viz., the dark night became bright from dusk till morning, and it appeared to the inhabitants that the adjacent parts of the globe were illuminated, and a large body of fire moved over the town and remained in the south–east; all the people rose from their beds, for they thought it was day, it (the light) continued so eastward along the sea.'*

338 Is this mention of holly somehow, besides meaning *peace*, also a pun on thorn? More curiously, is there a connection between the kiss of heaven and earth and the kisses Gawain delivers to Sir Bertilak, whose very greenness has

often struck commentators as a sign of a vegetation spirit or earth spirit? If the poet is indeed alluding to a comet grazing Earth, then the whole idea comes perilously close to what Mike Baillie has suggested is the most likely scenario for the climate collapse from 537–540 A.D., the latter date being one traditionally associated with the death of King Arthur at the battle of Camlann. Baillie's re-opening of the case for the causes of the Black Death, based in part on ice–core evidence and dendrochronological analysis, suggests that there is more to be said for the proximity between a comet and earth than has been given credence in the last century.

The other possibility regarding the identity of Bertilak de Hautdesert is that 'hautdesert' might not be in England at all. Although it may mean 'high hermitage', it may still be a reference to the deserts of Syria and it is possible that Bertilak goes back, like St. George, to Al Khidr, an ancient mythic pre-Islamic figure who features in the Koran and who is sometimes called the 'Green Elijah' or 'Green Moses'. The riddles and bizarre behaviour of Al Khidr are closely reflected in some stories of Merlin.

On this possible connection with Merlin, August Hunt has pointed out that the English translation of the Arthurian *Vulgate* renders the name Bertholais as Bertelak and that Bertholais may in turn owe his name to Britaelis from Geoffrey of Monmouth's *History*. 'Significantly, Britaelis was Gorlois's servant whose form was assumed by none other than Merlin in the story of Ygerna's seduction by Uther. If Bertholais is Merlin, it may be significant that the *Life of St. Kentigern* has Lailoken/Myrddin/Merlin buried "not far from the green chapel where the brook Pausayl flows into the River Tweed." In other words, this southern "Green Chapel" may be a relocation of the Northern Merlin's supposed grave.'

If, however, the oriental origin of Bertilak is to be favoured then perhaps in some measure Lady Bertilak also leans in this direction. The poet may have used Mary Gipsy (or St. Mary Gyp or St. Mary the Egyptian) as his model for her behaviour. The cult of Mary Gipsy was brought back by the first Crusaders, along with that of St. George, and she became extremely popular because of her renowned tolerance for sins of the flesh. (She was said to have gained passage on the only ship she could find going to Egypt—where she wanted to become a hermit—by granting sexual favours to the entire crew.) Swarthy in countenance, she seems to have become conflated on the one hand with Mary, the mother of Jesus (resulting in a large number of Black Madonnas throughout Europe) and on the other hand, with some of the ancient goddesses whose names were similar to her own, The Morríghan and 'Maid' Marian (both sometimes said to have the black–face of Morris Dancers) being two of them. The Morríghan is often said to be the source of Morgan la Fay (the fertility story of her union with The Dagda at the ford paralleling that of Morgan and King Urien) while Morris dancing is sometimes connected to the story of Gawain whose pentangle has been thought to be evocative of a ritual sword dance where blades are interlocked in a pentagram with a cry of 'A Nut! A Nut!' 'Nut' in this instance is said to mean 'knot,' as in 'endless knot'. Jessie Weston (1957) reports that Cecil Sharp noted four varieties of sword dance in Yorkshire and that it was particularly prevalent in the North of England and Scotland. One figure of the *Giant's Dance* which featured Odin and Frau Frigg consisted of making a ring of swords around the neck of a boy. Alistair Moffat's contention that the motto of the Scottish border town of Hawick is 'the land of Odin, the land of the dead' is worth considering since Valhalla and Avalon are such closely related words; Morgan la Fay is frequently portrayed as the mother of Owain as well as the wife of Urien; both Odin and The Dagda possess cauldrons of plenty or of inspiration; the further north the closer Odin's name comes to Owain's until by the time the Shetlands and Faroes are reached, they are virtually identical.

339 This is far from inconceivable since Jewish–Christian dialogue in the late fourteenth century was occasionally very high profile. Shem Tov ben Isaac ben Shaprut, a Jewish physician living in Aragon, debated Cardinal Pedro de Luna (later Antipope Benedict XIII) on original sin and redemption in Pamplona on 26 December 1375, in the presence of select bishops and learned theologians. Shem Tov was the author of *The Touchstone* (c.1380–85), a polemical treatise of which nine manuscripts survive today. The Gospel of Matthew in the Hebrew language (rather than the normal Greek) is found interspersed among anti–Christian commentary in *The Touchstone's* 12th volume.

340 H. Clay Trumball, *The Salt Covenant*, Impact Christian Books, 1999, p.104

341 EI Condren, *The Numerical Universe of the Gawain-Pearl Poet: Beyond Phi*, University Press of Florida 2002 p.43. Condren also points out that the four poems themselves have a chiastic design: the central two, *Purity* and *Patience*, focusing on the Old Testament and having an overall four–line structure while the first and last are 'contemporary' era and both have 101 stanzas.

342 Oui, *l'armour* is French for love. But I could never quite believe that the *Pearl* poet would see a woeful pun like that as sufficient reason to play with a link between *armour* and *kiss*. In fact, I'm glad I suspected it would have put him off and so went looking for another reason.

343 From its chiastic positioning with the 202–word sentence at the beginning of the epistle—and the opening words there referring to God's sustaining of his creation, this discussion of the armour is also meant to reflect the same Pauline thought. However it also throws light on Line 1596 of *Sir Gawain and the Green Knight* where the man and his hundred hounds are mentioned. This line is also Line 5151 of the whole manuscript—corresponding to the 101st triangular number. It is here that Bertilak pierces the armour of the boar, suggesting according to Barrett in *Against All England* a parallel vulnerability in Gawain's own 'spiritual armour' as he is tempted by Lady Bertilak.

344 Traver points out in a footnote, as previously mentioned, that the Midrashic story of the fight between the virtues can only have arisen, according to Professor Barton, after Aramaic supplanted Hebrew, since it interprets Psalm 85 on the basis that there are two words of double meaning: 'meet' means 'fight' and 'kiss' is taken as 'arm one's self'. This brief comment however does not constitute solid exegesis.

345 JA Burrow, *A Reading of Sir Gawain and the Green Knight*, Routledge & Kegan Paul, 1965, pg 39

346 Using the gematria calculator at BibleWheel.com (http://www.biblewheel. com/gr/GR_Database.asp)

347 In gematria, the *kollel* or 'one more' may sometimes be added to the numerical value of the word. According to David Patterson, the addition of the *kollel* is not simply for the word itself but for the silence within the word. (David Patterson, *Wrestling with the Angel: Towards a Jewish Understanding of the Nazi Assault on the Name*, Paragon House 2006, p. 194)

348 In fact the next nearest number is over 370 away so. Bearing on the choice of 77792 as close enough to indicate 77777 while being divisible by exceedingly significant numbers is that it is also divisible by 52 which is perhaps symbolic of time since 52 is the number of weeks in a year.

349 Let's just take a mathematical tour starting with the word 'armour' in Line 567 of *Sir Gawain and the Green Knight*. Since this number is divisible by 7, it may be a dourad and thus part of a Wheel of Fate formulation. If this is indeed the case,

then the full wheel would culminate on Line 1782, where the lady of Hautdesert accuses Gawain of having a lover. This is moving towards a critical changepoint but it certainly isn't the spot itself. However, if we inspect the *tau* of this line, we discover it is 1101 and there the plot has turned to the pledging of a covenant between Bertilak and Gawain over the exchange–of–winnings game.

Thus the armour that we began with is linked through the application of the Wheel of Fate and the golden ratio to a covenant. Here is a link to a destiny concept long associated in Hebrew culture with that of the armour–bearer: a covenant. The formal beverage–sealed covenant undertaken by Bertilak and Gawain is as weighty in obligation as any blood–brother pact which characterised the armour–bearer covenant. (In addition, of course, to the normal obligations of guest and host—a hospitality which, in the Middle East, was covered by a 'threshold covenant'.) Finding the *tau* of 1101, we come to Line 680, which mentions *'noȝt'* —zero or nought. This is 1211 lines from the beginning of *Patience*, itself a *'noȝt'*. 1211 lines is so close to 1212 lines of *Pearl* that perhaps this is a necessary hint that *Pearl* has zeroes sprinkled through it.

350 Louise Bishop makes the interesting observation that the idea that women and the lower orders of society could not read during medieval times is probably far from accurate. After all Lady Bertilak (in Line 1515 of *Sir Gawain and the Green Knight*) in fact casually refers to what she's read about Gawain. Women, according to Bishop, could not read Latin but they seem to have been well-versed in the vernacular—the idea of the home school may be far from modern. Children seemed to have learned to read at their mother's knee. Thousands of fragments of extremely well–thumbed medical books have been found that were obviously for women and by women. The element common to all of these texts (fragmentary because, unlike *Cotton Nero A.x*, they were in regular use) is that they are all in the vernacular. Arundel's interdiction, in Bishop's opinion, therefore may have been directed not just towards the Lollards but also towards ensuring that women and the lower classes did not get access to theological learning which was previously only available in Latin.

351 No virtue was invested with fame or worldly power for most medieval writers. They delighted in describing kings and wealthy bishops suddenly brought low by the hand of God through the agency of Dame Fortune. 'For example, on the death of William the Conqueror his supporters deserted him to seek out the new king, and his servants stripped the bed linen out from under the royal

corpse. In the Middle Ages, holiness, rather than fame or wealth, was the way to obtain status.' Jeffrey Burton Russell, *A History of Medieval Christianity- Prophecy and Order*, Thomas Y. Crowell Company, 1968

352 A question often asked is: who is the Green Knight? Is there a real Green Knight, Ralph de Holmes? Or a Green Count based on Adameus VI, Count of Savoy? Or is the Green Knight, to whom many 'fairy formulae' are attached, a vegetation spirit or an old god of the earth? I don't think the latter is possible from a poet with this particular set of sensibilities and with such a deep knowledge of the subtleties of various other religions. If I may indulge an intuition, may I suggest that the Green Knight represents, at least on one level, Time? And— all the deadly serious theological aspects of justice and mercy I've noted notwithstanding—nonetheless the critical scene between Bertilak and Gawain at the Green Chapel is perhaps also built around a quite dreadful pun: was it all about the nick of Time?

353 Mentioned by Plato in *Theaetetus*.

354 Jonathan Olsen, private correspondence, remarks on the use of $^{62}/_{101}$ to divide sections of *Paradise Lost*. Milton clearly knew the significance of truth in mathematical terms as indicated in his *Areopagitica*, which may contain an allusion to both 17 and Pythagoreanism:

Truth indeed came once into the world with her divine Master, and was a perfect shape most glorious to look on: but when he ascended, and his Apostles after Him were laid asleep, then strait arose a wicked race of deceivers, who as that story goes of the Ægyptian Typhon with his conspirators, how they dealt with the good Osiris, took the virgin Truth, hewd her lovely form into a thousand peeces, and scatter'd them to the four winds. From that time ever since, the sad friends of Truth, such as durst appear, imitating the carefull search that Isis made for the mangl'd body of Osiris, went up and down gathering up limb by limb still as they could find them. We have not yet found them all, Lords and Commons, nor ever shall doe, till her Masters second comming; he shall bring together every joynt and member, and shall mould them into an immortall feature of lovelines and perfection. Suffer not these licencing prohibitions to stand at every place of opportunity forbidding and disturbing them that continue seeking, that continue to do our obsequies to the torn body of our martyr'd Saint. ... They are the troublers, they are the dividers of unity, who neglect and permit not others to unite those dissever'd peeces which are yet wanting to the body of Truth. To be still searching what we know not, by what we know, still closing up truth to truth as we find it (for all her body is homogeneal, and proportionall), this is the golden rule in Theology as well as in Arithmetick, and makes up the best harmony in a Church; not the forc't and outward union of cold, and neutrall, and inwardly divided minds.

The major source for the myth of Osiris was the high priest of Delphi—Plutarch—who related that it was because of the Egyptian god's dismemberment into 17 pieces that the Pythagoreans loathed that number. This explanation seems extremely odd and, as far as I am concerned, should be treated with the utmost caution, though it rarely is. In terms of a modern parallel, would we expect the Dalai Lama to be the most trustworthy source for the basic tenets of Hinduism? It's not a question of integrity regarding Plutarch's statement, it's a question of viewpoint.

The Pythagorean Brotherhood worshipped Pythagoras as an avatar of Python Apollo, after whom he was named. Furthermore they were notorious for being atheists which, in those days, meant that like the Jews and Christians (who were also considered atheists) they did not worship a multiplicity of gods. They stuck to one. So, while it is not impossible that they should so reverence Osiris that they classed 17 as a repulsive number, it seems very bizarre.

Given the rise of Christianity in the first century and its fierce dedication to using the number structurally as often as possible (John even opened his gospel with 17 words), it seems more likely that the real reason had to do with this radical sect emerging out of Judaism.

355 By far the biggest difficulty with pursuing an investigation into mathematical grammar is that almost every researcher has worked independently and thus has invented a unique name for the phenomenon. When so few researchers share the same name for the same thing, it is very hard to find what others have uncovered. Nonetheless, the casual way Røstvig drops mathematical tidbits about various medieval poets as a prelude to her own arguments suggests she is drawing on a wide body of European scholarship not generally or widely available in English.

356 In *Europäische Literatur und Lateinisches Mittelalter*, Bern 1948, and *Neue Wege*, Utrecht 1950, respectively.

357 Maren Sofie Røstvig, *The Hidden Sense*, Norwegian Studies in English, 1963. p15.

358 Critics who disdain the very idea of numerical literary style, suggesting that no poet would bother with such a cramp on their creativity are really imposing their own cultural and personal biases on the literature of the past. 'Can anyone do this? Even if they can, why on earth would anyone attempt this?' is the disbelieving refrain. If I may venture a few possible answers: *Because it was the way things were done since at least the time of Plato and beyond that into the time when the book of Genesis was first written down. Because, like Everest, it's there to be conquered.* Not only is it an irresistible challenge, but it's not a particularly hard one. Long before I was sure numerical literary style existed, when I first heard rumours of it, I was already entranced with the concept. I began to write in the style just to see how difficult it was. Not only is it easy to fall into a rhythm with it but, with modern computers, it's extraordinarily simple in terms of word count and character count. Syllable count is a different matter. I can imagine, as Menken points out, that if you were taught how to do this from childhood it would not be difficult to master. I certainly don't try for the stunning architecture of the *Pearl* poet but I do try to use more than one metaphor at a time. This does create problems for an editor, however, so you really have to have someone in sympathy with your aims. I have now written 4 books (2 fiction, 2 non-fiction) and dozens of articles using numerical literary style. My children's fantasy *Many-Coloured Realm* has 111111 words in total, 1111 in the prologue and 111 words in the back cover blurb. 'Love, faith and the theory of relativity' might sum it up in one way but then so might 'Truth, justice and the theory of relativity'. No prizes for guessing where the concept of the run-of-ones comes from.

This book is 77777 words long in the main body (excluding the Index at the back and the Contents at the front.) These footnotes are 33333 words long. Not quite totaling 111111 words but close enough for the intention to be obvious.

359 As mentioned previously, 6666 turns up, but so does 153 and 170. Since 6666 has factors 2, 3, 6, 11, 22, 101 and 1111 and since 153 appears as well, I was suspicious enough to check out the series of poems Howlett indicated, speculating that 490 would be there as well. It was.

Paradise Lost has structural division of 798 and 1290. The latter might be from the last verse of the Old Testament's Book of Daniel, but in my opinion it is more likely that these are both chosen because 798 is the golden section of 1290 and because $^{62}/_{101} \times 798$ is 490. Milton, for whom the Music of the Spheres was a lifelong preoccupation, could hardly have failed to use 101 in this important way.

www.ingramcontent.com/pod-product-compliance
Lightning Source LLC
Chambersburg PA
CBHW031936090426
42811CB00002B/198